METAPHOR and its MOORINGS

M. Elaine Botha

METAPHOR and its MOORINGS

Studies in the grounding of metaphorical meaning

PETER LANG
Bern · Berlin · Bruxelles · Frankfurt am Main · New York · Oxford · Wien

Bibliographic information published by Die Deutsche Bibliothek
Die Deutsche Bibliothek lists this publication in the Deutsche Nationalbibliografie; detailed bibliographic data is available on the Internet at ‹http://dnb.ddb.de›.

British Library Cataloguing-in-Publication Data:
A catalogue record for this book is available from The British Library, Great Britain

Library of Congress Cataloging-in-Publication Data

Botha, M. Elaine.
Metaphor and its moorings : studies in the grounding of metaphorical meaning / M. Elaine Botha.
p. cm.
Includes bibliographical references.
ISBN-13: 978-3-03-910457-4 (alk. paper) 1. Metaphor. 2. Knowledge, Theory of. 3. Religion and science. 4. Science--Language. 5. Metaphor in the Bible. I. Title.
P301.5.M48B68 2007
401'.4--dc22
2007032095

Cover design: Vincent Allers, Renaissance Studio, Mossel Bay, South Africa

ISBN 978-3-03910-457-4

Hochfeldstrasse 32, Postfach 746, CH-3000 Bern 9, Switzerland
info@peterlang.com, www.peterlang.com, www.peterlang.net

TABLE OF CONTENTS

Acknowledgements

Permission was received from the following journals to republish articles in revised form.

1993 Michael Faraday en die letterlike lesing van die twee boeke van God: 'n gevallestudie oor. die verhouding van geloof en wetenskap. *Koers,* 58(2), 1993,141–168. (*Michael Faraday and the literal reading of the two books of God*)

1993 The mediating role of metaphor and analogy in the relationship between science and religion. *Tydskrif vir Christelike Wetenskap*, jrg. 29, 4e kwartaal, 1993, 60–73.

2006. Metaphor, embodiment and fiduciary beliefs in science. *Tydskrif vir Christelike Wetenskap* Festschrift D.F.M. Strauss. ISSN 1031–1116, p. 17–36.

2006 Metaphors, domains and embodiment *Koers* Annette Combrinck Festschrift. 28 pages.

2004 Clusters, roots and hierarchies of metaphors in Scripture and the quest for Christian Scholarship *Koers* 69, (3), 2004 28p.

The chapter "Metaphorical Models and Scientific realism" is a revised version of Metaphorical models and scientific realism. *Suid-Afrikaanse Tydskrif vir Wysbegeerte,* 5:3, Augustus, p.83–87. Also published in: Fine, A. & Machamer, P. *Proceedings of the 1986*

Biennial meeting of the Philosophy of Science Association, vol.1, East Lansing, Mich: PSA. 1986.

Foreword

The work embodied in this book represents the main thrust of my research on metaphor over an extended period of time. Tracing the sources of my interest remains a difficult task. Too many people and factors have played a role in the development of these ideas to be able to recognize and thank them individually. The intrigue that metaphor has held for me over a long period is deeply embedded in a religious quest to understand how one actually traces the constitutive presence of religious convictions in scientific theorizing. I am indebted to many colleagues, who have grappled with similar philosophical issues and who have provided guidance along the way.

Research for and preparation of many aspects of this book was supported by three Senior Research Grants of the Human Sciences Research Council in South Africa between 1975 and 1991 and a Standard Research Grant from the Social Sciences and Humanities Research Council in Canada (2001–2005). Much of the research was made possible through leave granted by the Potchefstroom University and the Senate Committee for Reformational Scholarship when I was professor of Philosophy at the Potchefstroom University for Christian Higher Education (1969–1995) in South Africa. I also acknowledge with gratitude a Research Fellowship of the Pascal Centre (1991) and a grant from the General Research Fund of the President, Redeemer University College, Ancaster, Ontario (2005). During this time, I was privileged to learn from many colleagues and researchers. I am grateful for the time was able to spend at the University of Notre Dame under the tutelage of Ernan McMullin and for the privilege of becoming more closely acquainted with the person and work of Mary B. Hesse who inspired me through her writings, insights and friendship. I would also like to thank Per Aage Brandt from the Center for Semiotics at Aarhus University in Denmark for the privilege and opportunity of spending some time teaching and interacting at the Center in the winter term of 2005.

I have been able to teach courses on various aspects of metaphor in three countries: South Africa, Canada and Denmark. Numerous students accompanied me in this quest. I owe them a debt of gratitude for patience, critical responses, solid research and encouragement. Students at The Institute for Christian Studies in Toronto, Redeemer University College in Ancaster, Ontario and the Center for Semiotics in Aarhus, Denmark most certainly provided the stimulus to continue with this research.

Many others have substantially contributed to this work through reading versions of the manuscript and copyediting. I would especially like to thank Roberta Podbielancik for her hard work and kind assistance in this respect, and the members of the SSHRC Standard Research Grant supported Workshop on Metaphor in 2004: Lori Kanitz, Svetla-Pacheva-Karabova, Anna Alouisa Moser, Deborah Bowen, Jo Kuyvenhoven, Craig Bartholomew, Al Wolters, Leo Mos and Chris Peet for their comments and critique. I also thank Derek Schuurman, Wayne Norman and the kind Information Technology Staff of Redeemer University College for their assistance with technical matters.

Introduction

The demise of the objectivist paradigm of knowledge, new embodiment theories of metaphor and empirical research concerning metaphor comprehension in a wide variety of disciplines, have brought about changes in the understanding of the nature of knowledge, reference, truth, meaning, reality and language and its relationship to the world. These theories have also provided new incentives to articulate views of embodiment that can do justice to both the multiplicity of meanings generated and discovered through metaphor and the multi-vocality and multi-facetedness of the possible worlds to which metaphors refer. As the result of these and other developments, so called non-epistemic factors such as socio-cultural, religious, metaphysical and ideological, formerly regarded as epistemically 'out of bounds' in the realm of science, are now seen as constitutive of conceptual meaning, cognition and conceptual changes in both ordinary experience and theoretical endeavours.

Any foray into the field of metaphor and its moorings faces the challenge to clearly focus on one set of issues without necessarily opening a hornet's nest of other closely related problems. Discussions of metaphor inevitably lead to a labyrinthine trail of often-unresolved issues in a vast interdisciplinary terrain. The discussions of these issues display similar characteristics to the phenomenon they are trying to grasp – they are wide-ranging, characterized by multivocity and hard to pin down. The complex discussions fan out over a wide-ranging area from traditional rhetoric and literature studies to cognitive linguistics, cognitive psychology, cognitive science, religion and theology. A highly theoretical vocabulary has developed which addresses issues starting from the conventional, ordinary language distinctions between literal and metaphorical language and meaning to detailed empirical analyses of meaning transfer, cross-domain mapping, conceptual blending, knowledge acquisition and ontological issues. Even though many of these discussions constitute unique discourses with their own technical vocabulary they have common themes which include the relationship between objectivism and relativism, the mind-body polarity, alternative views concerning cognition, reference and language, meaning, interpretation, reality, truth, etc. Theories about the pivotal role of metaphor in these domains abound, as do diverse accounts of how metaphorical mappings and analogical meaning transfer take place on the level of perception, concept formation, conceptual integration, language, communication, cognitive processes in general and scientific cognition specifically. Central to all of these accounts is the issue of the mediating, hermeneutical and semiotic function of metaphor and questions concerning the way metaphorical meaning is generated, conditioned and constrained.

Many have attempted to give an account of the nature of analogical information processing within metaphors and metaphorical meaning transformation across diverse contextual and semantic domains. Hesse (1988a:317–340), Rothbart (1988:377–399), Johnson (1987), Lakoff and Johnson (1999), Kittay (1987), MacCormac (1985), Kövecses (2002; 2005) et al have recently provided us with theories which have supplemented, modified, accommodated and clarified existing accounts of analogical understanding, and the mechanisms of meaning shift which take place when concepts are 'displaced' (Schön, 1963) through changes in theory. Overall, these accounts are related to and dependent upon general metaphor theories of which so many are in circulation that various classifications have already been made. Beardsley (1967:285–286), Scheffler (1979), Kittay (1987:23), Rothbart (1988), Kövecses (2002; 2005) have all proposed such theories. Significant new avenues of exploration are the so-called 'embodiment'-, 'domain'- and 'mental space' theories. Lakoff and Johnson and Rohrer et al's introduction of 'embodiment' and image schemata as the basis of cognition accommodates factors previously regarded as 'subjective' or non-epistemic in the act (interaction or enactment) of cognition. Their recognition of diverse but related domains of experience present in the embodied cognitive act also opens up new avenues for the exploration of the constitutive role of religious convictions within scientific cognition.

Traditional understandings of the double language thesis with its moorings in the objectivist paradigm of cognition and representational-computational view of mind are being challenged and redirected by the introduction of new notions of embodiment. This has forced scholars to revise theories of metaphor to accommodate the questions raised by embodiment theorists from both empirical and philosophical angles. Weighty evidence in recent empirical research in cognitive semantics and cognitive linguistics demonstrate that metaphors are more than mere lingual phenomena and are based in 'experiential gestalts' expressing embodied human understanding and empathetic interaction between human beings and between human beings and reality. This has led to postulating different positions concerning the nature of the embodiment, truth, cognition and reference. Lakoff and Johnson (1980; 1999) and others have proposed the notion of 'conceptual metaphor'. Sweetser (1990) and Brandt (2000) in turn have worked in 'domain theory' and Fauconnier (1994) has proposed the theory of 'mental spaces'.

I propose to build critically on these insights by introducing a new angle on current discussions concerning conceptual metaphor and its basis in human embodiment. Can the multiplicity of possible meanings which metaphors generate be grounded in the human mind, the body or in the nature of the world? Perhaps the most fertile and productive line of argument localizes metaphor in the embodied inter-acting human being, creating, discovering and opening up new meanings in this process. A view of embodiment will be presented which grounds conceptual metaphors and the wide spectrum of possible meanings and worlds in the ineradicably relational nature of human embodiment which includes at least the following four levels: The relationship of the whole human being to herself, to others, to the environment, culture and history and to that which is regarded as ultimate, god, spiritual, God, etc.

Arguments which claim that faith intrinsically influences scientific theorizing often neglect to show *how* this is true. Proponents of this position often argue that it might be possible to demonstrate this claim in the social sciences and humanities, but that it is impossible in the realm of natural science. It is also often argued that such influence might be demonstrated in historical periods when both culture and science was strongly influenced by religion, such as medieval times and the seventeenth century, but that the demonstration of the claim is impossible in highly secularized periods when the influence of religion is not as apparent. I propose here that religious convictions actually do influence both social and natural sciences and that religious factors are at work in scientific theorizing even in apparently secular settings. In order to demonstrate this a redefinition of the term religion is argued for, the classical double language thesis (the traditional distinction between 'literal and metaphorical' language) is critically assessed and a methodology of metaphorical hermeneutics is introduced. I shall be working with a reformulation of the notion of religion as proposed by Clouser (1991, chapter 2). The significance of this reformulation is that religion is now not necessarily related to gods/God, worship, liturgy, moral and ethical practices, but to that which is regarded as divine and therefore non-dependent. This notion of religion provides a significant basis for distinguishing between the deeply religious function of root-metaphors on the one hand and the subsidiary, tacit, fiduciary role of some controlling (analogical) beliefs on the other hand. The proposed methodology shows metaphor and its underlying analogical structure to be a significant key to the understanding of the 'metaphorical' nature of both reality and cognition. It acknowledges the embodied nature of human understanding, proposes the distinction of a number of universal domains of meaning and facilitates the understanding of the mediating function of metaphor between deeply held metaphysical beliefs and theorizing. This will show that the restriction of metaphorical meaning to primarily a linguistic notion is inadequate and requires deepening and expansion.

The still controversial claim that 'all language is metaphorical' and that metaphors have cognitive content, forms the basis for the argument that metaphorical and analogical language mediates between religiously determined views of reality and the actual theories developed based on such metaphors. The groundbreaking work of Kuhn with its recognition of the role of paradigms, metaphors and analogies in science, has been appropriated by some scholars in defense of claims that one's world view and more specifically theoretical view of reality, are the vehicles which convey religious assumptions into theoretical work. Although I doubt whether Kuhn's views actually substantiate this, I believe there is merit in pursuing his proposals in order to trace the constitutive role of metaphor and analogy in theorizing. But more is required in order to show that religious convictions are constitutive of theorizing. I argue that both pre-theoretical and theoretical world views are often deeply rooted in root-metaphors of a religious nature which permeate theories in constitutive and decisive ways.

The changes in the understanding of the nature of language and specifically metaphorical language have also had significant consequences for the interpretation of religious language and more specifically religious metaphors and the way one views the influence or presence of religious convictions in scientific theorizing. Moreover, in some of the primary literature on metaphor the issue of 'spirituality' surfaces and is appealed to as a basis for embodied metaphorical meaning. These proposals need serious and critical attention and the consequences of the arguments presented for the notions of scientific realism and spirituality require further exploration.

This project follows an interdisciplinary approach, drawing from recent developments in a diversity of academic fields. It also attempts to reformulate and refocus the traditional discussions of the relationship of science and religion in light of new understandings of both metaphor and embodiment. The project explores the significance and impact of these embodiment theories of metaphor on some traditional formulations of the relationship between science and religion and introduces a methodology based on the distinction of a number of universal domains of meaning discernable in both human embodied experience, cognition and reality. This will show that 'metaphor' needs to be distinguished from proper analogy and that literality is as much a problem as metaphoricity because literality is as much dependent upon classification and categorization of reality as is metaphorical language (Hesse, 1983:40). Having said this, the classical problems related to metaphoricity and literality are obviously not resolved, but moved to a different level. In this work, I propose the recognition of the contextual qualification of the distinction between literal and metaphorical language and the relativity of both literal and metaphorical meaning. I maintain that the possibility of stable meaning and communication is guaranteed through the introduction of the notion of 'proper analogy' which functions as the mooring for metaphorical meaning.

Perhaps the most significant findings with respect to metaphor have been that both literal and metaphorical language (the double language thesis) is contextually conditioned, that the distinction is relative and does not necessarily require a two-tier view of cognition or reality. The distinction is also ultimately related to that which is regarded as the acceptable, conventional classification or categorization of the pre-conceptual and experiential structures of reality at a specific historical point in time. Metaphor brings about the expansion and changes in such classifications and categorizations of reality. This holds for conceptual displacements in everyday language, meaning shifts in theoretical language and certainly also in changes of understanding concerning the nature of God and His revelation.

I suggest that the literal-metaphorical distinction needs to be replaced by a distinction between non-analogical and analogical meaning of domains and concepts. The non-analogical meaning is the original, primary, 'home base' or domain meaning that is characterized by a unique meaning nucleus. This is the way in which one of the multiplicities of original, irreducible semantic domains of reality presents itself. The intriguing characteristic of the non-analogical domains is the fact that they harbour the full gamut of potential references to the wide spectrum of meaning of our multi-ordinally structured world. When concepts are formed the meaning of these non-analogical concepts are expressed in a vast array of analogical moments which relate this original, non-analogical primary, primitive, univocal meaning to other semantic domains. Once a concept from such an original source domain is utilized in any other domain, it acquires a qualification which indicates its analogical use. The history of the differentiation of scientific disciplines rests on the process of exploring and opening up these analogical elements of original, irreducible domains of reality (Stafleu, 1978; 1979).

The fiduciary, certitudinal or confessional dimension of reality constitutes one such non-analogical or original semantic domain characterized by the nucleus of 'trust'. All entities in reality (facts, things, events, acts and relationships) passively or actively participate or function in this domain – so do theories and models. In any theory attempting to understand reality, there are always a number of assumptions or presuppositions which are accepted on face value and which act in a fiduciary manner, without being questioned. In the conceptual frameworks of such theories concepts are utilized which acquire a meaning from some original domain, e.g. force as a physical phenomenon. When the meaning of the term is unpacked, it becomes clear that it is imbued with a specific conceptual connotation. A good illustration is Hesse's (1988a:120) comparison between the difference in meaning of the concept 'negative charge' in 19th century atomism and electromagnetic theory. I argue that the difference in emphasis on a specific analogical moment in each one of these two theories play a significant role in the differences of meaning of the concepts utilized in the respective theories. Ultimately the decisive fiduciary role played by such an analogical moment ought to be taken into account when attempting to understand the basic difference between these two theories.

Amongst these analogical moments that are present in theoretical language, there is often a decisive, regulating and controlling analogy constitutive of the metaphorically constituted theory terms. The controlling analogy within a theory or model based on a metaphor, often functions as the fiduciary component of a theory. This proposal rests on the assumption that a fiduciary component is a significant domain of embodied human existence, cognition and theorizing. Therefore, it is constitutive of all acts of knowing that are obviously always embodied. These distinctions warrant the introduction of a clear demarcation between metaphor and analogy.

Metaphors are usually used in concrete references and refer to entities. Metaphors can be replaced by other metaphors. Yet, there are some metaphors which refer to some deeper ontological coherence of the diversity of reality and are therefore of a different kind. They cannot easily be replaced by other metaphors. Such metaphors are original, primitive or non-analogical. Hart (1984:156) calls these metaphors, which express analogical relationships between aspects or facets of reality, *modal*. Modal in this sense should be differentiated from notions of 'modality' used elsewhere in philosophy, logic and metaphor theory. It refers to aspects or facets of reality or to different ways in which reality functions. These proper analogies or ontological metaphors or unavoidable analogical structures play a significant role in the process of uncovering the unavoidable analogical moments that are pregnant expressions of the coherence and interrelatedness of original and irreducible domains in reality. This becomes apparent in the way metaphorical modeling functions in theorizing. Metaphors mediate access to these ontic analogies through the process of analogical concept formation. These concepts are closely related to the basic metaphor or root-metaphor operating in a discipline or theoretical perspective.

I will argue that metaphorical language and cognition reflects the intrinsically 'metaphorical' nature of the embodied human experiential and conceptual system and also correlates with the 'metaphorical' stratification of reality. I suggest that the methodology of metaphorical hermeneutics will show that metaphor and its underlying analogical structure are significant keys to the understanding of the 'metaphorical' nature of reality and cognition. The changes in the understanding of the nature of language and specifically metaphorical language have significant consequences for the interpretation of religious language and religious metaphors – a topic that will only be dealt with *en passant.*

For the Christian scholar all of this raises a number of questions: How does the metaphorical nature of religious and Biblical language relate to the metaphorical nature of everyday and theoretical language and how do they both in turn relate to growing insights into the metaphorical/analogical structure of the world we live in? In which ways does Biblical metaphorical language and religious language in general distinguish itself from metaphorical language used elsewhere in a diversity of fields? Does Scripture harbour central or root metaphors or a hierarchy of such metaphors that regulate the itineraries of meaning (Ricœur, 1980) of the full story line of Scripture and therefore dictate the reading and interpretation of the diverse texts that constitutes it? Which candidates are the most likely and most Biblically responsible root construals of the central reality of Christianity?

I shall argue that Biblical language and religious language is intrinsically certitudinal and confessional in nature and relates to the concrete, experiential relationship of the community of faith. One dimension of the life of the community of faith is scholarly and academic activities. When seeking 'justice-in-shalom' in this dimension of life, the scholar is called to work out the clams of her religious allegiance in the area of theorizing. It requires at least an exploration of the relationship of theoretical and other root-metaphors to the overall itinerary of meaning of the overarching Biblical story. In the final instance, the central question raised in the title of this work will be dealt with, which is tracing the consequences of the ontology introduced above, for the realism debate and the debate concerning the role of religious convictions in science. I argue that the recognition of original and irreducible domains of reality approximated by metaphorical language and models holds promise for the resolution of the interminable double-language thesis debate.

Part One

Metaphor and Embodiment

I The Knowledge of Our Body and the Body of Our Knowledge

Metaphor: Between a Positivist Rock and a Postmodern Soft Place

Postmodern challenges to modernist assumptions about rationality, truth, realism and reference have led to a fundamental rethinking of the cognitive bases of knowledge. Reflection on the nature and cognitive claims of metaphor has been central to this process. It has called into question the traditional Aristotelian view of metaphor with its literal truth paradigm and objectivist view of knowledge in which the human conceptual system is seen as essentially literal. In the traditional Aristotelian view, metaphor is seen as deviant and emotive use of words and meaning and truth claims are reduced to their literal paraphrase. Rooted both in the classical Aristotelian view and in the philosophical assumptions of modernity, the objectivist view of knowledge saw metaphors as words used in a context different from their original or proper sense and therefore emotive and aesthetic or inevitably deceptive, seductive and misleading.

The positivist treatment of metaphor that prevailed during the first half of the twentieth century is actually a version of the centuries-old empiricist critique of metaphor. The positivist's critique rests on two foundations (Johnson, 1981:17): The distinction between the alleged 'cognitive' and 'emotive' functions of language, and the attendant belief that scientific language can be reduced to a system of literal and verifiable sentences. These two tenets were conjoined with the earlier empiricist 'literal-truth' paradigm which argued that metaphor deviates from literal or cognitive discourse and, if it had any cognitive import at all, it would be expressible by literal statements of comparison. This tradition also fostered the so-called reductive theories concerning the nature of metaphor, i.e., the simile (substitution) and the comparison theories, which both attempt to reduce metaphorical expression to a substitution either of the one referent of the metaphor by the other or to a comparison of the two referents on the basis of literal similarities or analogies. Early challenges to the traditional view of metaphor were that of Richards, The *Philosophy of Rhetoric* (1936) and the seminal essays on metaphor by Black (1962; 1977; 1979) who questioned the double language thesis and the restriction of meaning to single words.

From a different angle, the views of representatives of the Historical Turn in philosophy, history and sociology of science like Polanyi, Toulmin, Kuhn and Nersessian, et al led to changes in the understanding of the nature of scientific cognition. In the bourgeoning empirical research on metaphor in Cognitive Science, Cognitive Psychology, Developmental Psychology, Cognitive Linguistics, Semantics, Semiotics and Categorization, there is a growing body of evidence which disproves many of the basic tenets of the traditional view of metaphor. The upshot of these developments is that many scholars have come to understand that it is not literal language but metaphor as a primary figure of speech that characterizes both language and thought in a most fundamental way. Moreover, it is now argued not only that human thought is metaphorical but that there is no unbridgeable gap between literal and metaphorical language. The literal-metaphorical distinction is relative, rather than absolute and literal and metaphorical meaning is always contextually determined. Perhaps the strongest move has been towards the understanding that the transfer of meaning, conceptual blending and imaginative creativity so characteristic of human language and thought, points to bodily-based experiential structures. Thus, embodiment has come to be recognized as the basis for metaphorical meaning.

With the advent of pluralism and perspectivalism and the accompanying recognition of the multivocity of concepts other questions arose: What are the constraints to our knowledge of reality? What guarantees stability of meaning and communication? Since no 'God's eye view' of reality is possible, how would we know when we have actually accessed 'the truth?' These developments placed the tension between 'objectivism' and 'relativism' centre stage to the discussions of metaphor. This tension between positivism and postmodernism is perhaps best demonstrated by Bernstein (1983) who captures the tension between objectivism and relativism in his book *Beyond Objectivism and Relativism. Science, Hermeneutics and Praxis* when he writes:

> [...] at the heart of the objectivist's vision, and what makes sense of his or her passion, is the belief that there is or must be some fixed, permanent constraints to which we can appeal and which are secure and stable. At its most profound level the relativist's message is that there are no such basic constraints except those that we invent or temporally (and temporarily) accept (1983:19).

Views that do acknowledge the existence of constraints Johnson (1989) argues, are related to a

> [...] disembodied understanding of objective rationality consisting of pure abstract logical relations and operations which are independent of subjective processes and sensori-motor experiences in the bodily organism [...] delineated by formal logic and held to be entirely value-neutral and free of emotional and imaginative dimensions [...](1989:110).

So the pendulum swings between the relativists' denial of the presence of constraints to the objectivist's ideal of disembodied reason as anchor for rationality. Closely related to the objectivist project is the 'common sense understanding' of the world as represented in the *representational-computational view of mind* (Varela, et al, 1991:7, 144, 218). One of the basic assumptions of representationalism is that there is a world with pre-given features mirrored by the internal representations of the mind (ibid. 1991:136). Cognitivism is the strongest version of the representationalist view of mind inaugurated by Descartes and Locke (ibid. 1991:135, 138), a view that has recently had to endure some devastating attacks from postmodern and other quarters. The central tool and guiding image of cognitivism is the digital computer with its connectionist approach to symbolic processing (ibid. 1991:7, 138). At the core of these discussions are questions concerning the nature of reality and reality depiction. The profound suspicion concerning the objectivist notion of knowledge, the postmodern appreciation for pluralism and relativism and the historical and sociological turn in philosophy of science all contributed to an appreciation that reality could be portrayed or represented in manifold ways. If there were only one representation of reality possible anti-realist positions would make no sense. If it were possible to know where and how representations actually correspond with reality, the whole discussion would be a non-issue. Both the empiricist tradition and postmodernism face the common challenge of providing an account for the grounding their respective positions.

Postmodernism, as Kvale (1995:19) has noted, gave rise to the loss of belief in an objective world and an appreciation for the decentralised and fragmented '[...] for heterogeneous language games and incommensurability of these language games'. A postmodern world is characterized by continual change of perspectives, with no underlying common frame of reference, but rather a manifold of changing horizons (ibid. 1995:21). On the surface, it seemed as if the widespread interest in metaphor with its ubiquitous and ambiguous meaning and rich sources of polysemy played directly into the hands of the postmodern worldview. It made it possible to describe and explain reality from a vast diversity of possible metaphorical points of view, without necessarily addressing the issue of limits or constraints, truth or reality depiction. The continental drift towards relativism brought about by these developments was somehow stabilized by closer scrutiny of exactly how meaning change and meaning transfer took place via metaphor. Empirical research was being focused on the actual experiential processes on which meaning was based. How did one know the 'meaning' of a term? What were the mechanisms of meanings shift? How are these meanings constrained by reality?

The work of Lakoff, Johnnson, Turner et al showed that human beings possessed 'image schematic structures' which somehow guided their 'experiential *gestalts*' of reality and that these were closely related to what came to be called 'conceptual metaphors' – basic human abilities to make sense of the world around them based on recurring patterns of experience. 'Image schemas' emerge from certain basic forms of sensori-motor activities and interactions and prove a pre-conceptual structure to our experience (Johnson, 1987:28, 29; Johnson, 1989:113, 115).

The quest for answers to these questions generated various theoretical avenues, such as 'domain theory' and 'mental space' theory. In these theoretical developments, the emphasis remained focused on empirical analyses. An important strand of the decisions concerning the grounding and claims of metaphorical knowledge was Rorty's (1981) rejection of the *mirror of nature* approach to knowledge and Hesse's (1983) persuasive arguments for the cognitive claims of metaphor.

Richard Rorty could be regarded as one of the most significant philosophers who addressed the flaws embedded in foundationalism, essentialism and the representationalist *mirror of nature* approach to knowledge. The full discussion of Rorty's critique of the Cartesian legacy provides too wide a context for the issues relevant to this chapter (cf. Malachowski, (ed.) 1990). I shall focus only on Rorty and Hesse's divergent views concerning the cognitive claims of metaphor (Rorty, 1985/6; Hesse, 1985/6). These include inter alia the question of literal meaning, the basis for meaning change brought about by metaphor and the rootedness of metaphorical meaning in some form of ontological constraints. An early significant article of Hesse (1983) and the Rorty (1987)-Hesse (1985/6) and Johnson (1991) discussions indicate some of the parameters of the discussion.

In his critique of the *mirror of nature* view of knowledge, Rorty unmasks some of the assumptions of the objectivist views of knowledge. He then argues that causal explanations of how one comes to hold certain views is a social practice of epistemic justification which is determined by the social group and dominant vocabulary of knowledge guiding the discussion. Shifts in these epistemic vocabularies are arbitrary shifts of optional metaphors, he argues (Rorty, 1989). Rorty follows Davidson (1978) and others in the belief that metaphors express nothing and that the only meaning associated with metaphor is its literal meaning (Hesse, 1987: 297). According to Rorty metaphors are nothing more than [...]'non-cognitive speech acts to be understood causally as 'noise' expressing nothing' (Hesse, 1987: 297). Metaphors are transitions to some ideal literal talk which will be arrived at once the tropical jungle has been cleared and the clear space of literal meaning becomes apparent. Hesse (1983; 1987) on the other hand, argues for the cognitive claims of metaphor and the 'myth of the literal'.

The Cognitive Claims of Metaphor Based on Irreducible Primary Relations

Applying the Wittgensteinian notion of 'family resemblances' to general terms Hesse argues that all applications of general terms, however apparently literal, depend on perceptions of similarities between their referents. In language acquisition, these perceptions of similarity lead to general terms which classify the furniture of the world according to the classifications of our culture (Hesse, 1987: 311). Although she does not develop a full-fledged ontology, she states the need for such a theory. She has appreciation for the work of Ricœur who embarked on such a project (Hesse, 1983: 38; Hesse, 1988a).

Earlier when Hesse (1983: 40, 41) developed her thesis concerning the cognitive claims of metaphor, she argued that all descriptive language implies a normative classification of the vast multiplicity of things in the world. These factual descriptions are value-, norm- and category-laden distinctions. In everyday life and science, this normative classification is constrained by criteria of prediction, test and self-correction. This is what determines 'facts', 'objective reality' or the 'literal' world. These views have consequences for what is regarded as cognitive. It requires a revised ontology, one which acknowledges that there are no hard and fast natural kinds like Aristotle thought. It requires a revised epistemology – there is no correspondence of statements to facts or reality. But in the final instance, there is no primary literal language with metaphorical language as a deviant form. It also requires a revised theory of truth. Truth is contextually determined by the constraints of the specific area we find ourselves in and truth values can be ascribed to metaphoric statements within a social network of conventions as wide as a specific vernacular or a particular religion or ideology. Metaphoric usage is concerned with re-made or possible worlds in two senses: change of viewpoints and the fact that it goes beyond factual description which has technical interest in prediction and control. The social network of meaning determines what is regarded as truth by a community. At this point, there seem to be important similarities between Hesse and Rorty's position. Both harbour elements of nominalism (Haack, 1987/88:301), yet Hesse also links her view of metaphorical meaning to the constraints provided by human physiology and our commerce with the world (1985/6:40) and describes her position as one of 'moderate realism'.

Metaphoric usage implies evaluative interpretations as well as descriptions. Reductionist theories of metaphor, Hesse argues, should be replaced by a network theory of meaning that acknowledges elements of the interaction view of metaphor as proposed by Black (1962). The double-language thesis, which she claims is based on the Aristotelian view of universals that assumes natural kinds and a 'correct description', needs to be replaced by a view that metaphorical meaning and meaning transfer rests on 'resemblances in certain respects' found in classes of objects that have been grouped together. These similarities and differences based on the way in which we classify things and meanings are going on in language all the time. Games, psychological types and schools of painting are examples of such groupings based on similarities. Metaphor is instrumental in creating the similarity, but she argues that the sense of similarity and difference on which the network view depends is primary and all pervasive – they '[...] are irreducible primary relations, prior even to application of the simplest predicate: they are *shown* and not *said*' (1983:33). These similarities are not of the Aristotelian natural kind type, but of the Wittgenstein family resemblance kind. It is the nature of these ontological constraints or 'irreducible primary relations' and the way they relate to social networks of meaning that Hesse emphasizes. She recognizes the need for a new understanding of an ontological framework and also clearly points in the direction where it can be sought. But she does not develop such a framework.

Currently there seem to be a number of contenders for such a framework, Lakoff and Johnson's 'embodiment thesis' being perhaps the most obvious one. I shall give a brief overview of this thesis with some reference to elaborations made by contemporaries and then proceed to the development of an alternative anthropology and ontology that could serve us well in opposition to the objectivist epistemology and provide the basis for an account of metaphorical meaning.

Grounding Knowledge in Embodiment

As the result of the demise of the objectivist paradigm of knowledge and its claims to certainty, epistemology moved into the soft place of relativism. The challenges posed by relativism called forth responses in philosophy, philosophy of science, literary theory and a number of other disciplines. North American philosophers, cognitive scientists and linguists have developed their own alternative non-objectivist stance in which knowledge is viewed as being the result of an ongoing interpretation that emerges from our embodied capacities of understanding. The 'embodiment thesis' has been proposed as a solution to some of the epistemological problems resulting from the demise of the Cartesian and Enlightenment objectivist epistemological legacy.

Significant contributions to this exploration have been made by Lakoff and Johnson, Turner, Rohrer, Shanon, et al. These proposals lead to a different explanatory basis for metaphorical meaning. They are of interest within the context of this study insofar as they address the issue of the grounding of meaning and more specifically metaphorical meaning. Different options are pursued by those theoreticians who explore some form of 'grounding' of meaning. Of these options the embodiment thesis developed by Lakoff and Johnson is certainly one of the most promising proposals. Rohrer (2005) says it was developed as a potential solution to the issue of directionality of metaphorical meaning. The body would provide the constraints on metaphorical mappings and the direction of the constraints would be from the concrete bodily structure to the more abstract domains.

In many respects, the views in which knowledge was seen as 'embodied' were not new. There were already numerous precursors to current views on 'embodiment' especially amongst Continental philosophers and more specifically Phenomenology (Edmund Husserl, Maurice Merleau-Ponty) Michael Polanyi and others. All these developments pointed in the direction of a far more dynamic and 'full bodied' understanding of the nature of knowledge than was the case in the objectivist tradition. Continental philosophers continued to produce detailed discussions that showed that knowledge '[...] depends on being in a world that is inseparable from our bodies, our language, and our social history – in short from our *embodiment*' (Varela et al 1991:149). This was a position much disputed by linguistic philosophy and logical positivism with its 'received view of scientific theories' (Suppe, 1974:3–233). This objectivist epistemology in association with what later came to be called the *cognitive-representationalist view of mind*, sees knowledge as static representations of a pre-given structure of the world. Developments in conceptual metaphor theory and embodiment theses have shown this approach to be fundamentally flawed.

Johnson (1991:7) agrees with Rorty that the foundationalist theory of knowledge needs to be rejected, as should the *mirror of nature* view of knowledge which '[...] consists of a fixed relation between either a mental *state* (=representation) or a descriptive *state*ment and a corresponding fixed *state* of affairs in the world' but then adds that it also requires the rejection of the attendant objectivist assumptions and the recognition that knowledge is ordered and contextualized (1991:15). Rorty's relativistic solution to the problems posed by objectivism and the *mirror of nature* position is attributed to the fact that he '[...] has an inadequate view of meaning and metaphor that misses the way in which semantic structure and semantic change have a grounding in patterns within our bodily experience' (1991:6). Johnson argues that Rorty's position makes the whole process of knowledge formation '[...] a non-rational result of sheer contingencies' (1991:5). He suggests bringing the body back into knowing and argues that this does not lead to extreme relativism, because our knowing interactions are constrained in certain ways by 'image-schematic patterns'. Such structures, Johnson argues, should be seen as '[...] recurring overlapping patterns in the ongoing flow of our experience of a world. These are the imaginative contours of our experiential interactions' (1991:15). In Johnson's view of embodiment, there is a strong emphasis on the non-propositional and pre-conceptual structures on which metaphorical meaning rests. The embodiment thesis is shared by George Lakoff and others and is also argued in publications that Lakoff and Johnson co-author, e.g. *Philosophy in the Flesh. The Embodied Mind and its Challenge to Western Thought* (1999:16–44). Hesse, too, ultimately appeals to the fact that the patterns that constrain meaning are those embedded in human culture and the physiological interaction with the world (Hesse, 1985/6:39).

Knowing Through the Body

Johnson (1989:109–118 and 1991:3–18) states that embodiment is the locus of our experience and specifies embodiment more precisely as the patterns that emerge through the sensori-motor activities of human beings (1991:8). This has to do with spatial and temporal orientation, bodily movement, perceptual focus and manipulation of objects to accomplish certain purposes. In the use of these sensori-motor skills, image schematic structures emerge and develop which provide the form, order and relative determinateness in human experience of the world. These image-schematic structures are also at work in the more abstract realms of understanding and reasoning '[…] from the most primitive and mundane unreflective acts of perception and motor activity all the way up to abstract reasoning and argument' (1991:12). All of this Johnson claims, is knowing. The image-schematic patterns are what Johnson calls 'imaginative structures'. They are the imaginative contours of our experiential interactions that are employed in order to have ordered, recognizable representations and experience. On the basis of his thesis that 'knowing through the body' can move us beyond the unproductive epistemological impasse, Johnson develops a very rich and nuanced non-foundationalist concept of 'knowledge'. He argues that we need to overcome our deeply rooted notion that knowledge is static, propositional and sentential (1991:3).

According to Johnson the problem with this type of epistemology is the fact that it works from the assumption that all knowledge of the 'knowing that' sort is sentential. He suggests we need start again, but this time from the phrase 'knowing through the body.' His epistemology sees knowledge as an activity in which environment and organism interact. He follows Patricia Churchland and others who argue that knowledge is a mode of activity in which the continuity of our bodily interactions with more abstractive modes of thought is the '[…] key to a unified view of knowledge as a means of transforming our experience' (1991:6). Following Dewey's emphasis on a conception of knowledge as an activity and knowing as an organism-environment interaction, Johnson argues that the knower is '[...] already engaged with their surroundings in an ongoing process of mutual interaction, adjustment and transformation'. He notes: 'Both environment and organisms are what they are at a given moment only in relation to each other and only as the result of a continual process of constructive interaction' (1991:7). This, Johnson argues, takes place on the basis of bodily experience based in emerging image-schematic structures of cognition (1991:6) which constrain metaphorical meaning. He defines *image schemata* as

> [...] structures in our *understanding and experience* of the world and, as such, are not ordinarily part of our *self-reflective* awareness, though they are part of our awareness. They can properly be called structures of understanding because they are patterns in terms of which we 'have a world' – which is what is meant by 'understanding in the broadest sense.

According to Rohrer's (2005), rendering of Johnson's view such image schemas are pre-conceptual embodied structures of meaning in at least two different senses: Developmentally they are prior to conceptual thinking, at least in so far as conceptual structure is accessible by means of language and they are pre-conceptual in that they underlay multiple different metaphors. The example Rohrer refers to is the extension of directly emergent spatial experiences to characterize non-spatial experiences. The notion of understanding as 'a way of being in' or 'having' a world emphasizes the interactive character of understanding and meaning is developed in opposition to what Johnson regards as the the erroneous objectivist claim that

> [...] only a viewpoint that transcends human embodiment, cultural embeddedness, imaginative understanding, and location within historically evolving traditions can guarantee the possibility of objectivity (Johnson, 1987:175).

Against this erroneous objectivist claim Johnson (1987) and Lakoff and Johnson (1988) develop their experientialist approach to cognition. In this approach Johnson argues for a notion of objectivity on the basis of shared '[...] image schematic-structures and their metaphorical projections (that) have a shared, public character which gives them a central role in the objectivity of meaning'. Johnson uses 'objectivity' in the sense that there '[...] are meaning *gestalts* connected to structures of bodily experience that we can all share' (1987:175). Meaning includes '[...] patterns of embodied experience and pre-conceptual structures of our sensibility (i.e., our mode of perception, or orienting ourselves, and of interacting with other objects, events, or persons)' (ibid. 1987:14).

The embodiment thesis and conceptual metaphor theory provides significant solutions to the questions raised by the objectivist and literalist view of knowledge. However, a number of issues require further elaboration. The role played by the stratification of reality in the process of cognition is one such issue. A second issue is the ontic grounding of the multiplicity of meanings embedded in the analogical concepts of the disciplines. Both of these issues will be addressed in Part II.

It is not only Lakoff and Johnson that are seeking ways to escape the objectivist dilemma. In the next section I shall briefly discuss three other positions in which non-objectivist alternatives to cognitivism are proposed.

Non-objectivist Alternatives to Cognitivism

Being and Acting in the World

The philosopher of science Hacking's theory concerning the nature of scientific theorizing is an example of an alternative to the traditional cognitivist view of knowledge. In his *Representing and Intervening Introductory Topics in the Philosophy of Natural Science* he writes: 'Anti-realism makes no sense when only one kind of representation is around' (1983:139). Hacking proceeds to argue for a 'homo depicter' theory of representation, one that distinguishes between the image formed of the world that is to be depicted and the linguistic articulation of this image. Hacking states that the line of argument in which one asks whether a scientific theory is true to reality is fruitless. He prefers to approach the issue of scientific realism from the angle of action. What happens in a new theory is not only a new way of thinking about the world, but a new way of acting in the world. So he replaces thought with action and theory with experimental intervention. Clearly the emphasis is on the dynamic nature of theorizing, an emphasis also shared in the views of Varela, Thompson and Rosch. Obviously action entails thought but it also brings a dynamic embodied perspective into play which broadens the scope of what knowledge incorporates. Action incorporates the embodied knower in the active process of knowledge formation. A similar position is developed by Varela et al.

Creative, Dynamic, Interactive and Context Dependent Know How

Varela et al develop their non-objectivist alternative to cognitivism as '[...] the view that knowledge is the result of an ongoing interpretation that emerges from our capacities of understanding' (1991:149). Their views correlate with an important shift in Cognitive Science: A move '[...] away from the notion of an independent and extrinsic world to a world inseparable from the structure of these processes of self-modification' which are in turn '[...] rooted in the structures of our biological embodiment, but are lived and experienced within the domain of consensual action and cultural history' (Varela, et al, 1991:139). These capacities are seen as being '[...]the manner of 'having a world'. (ibid. 1991:150). They argue that Continental philosophers have continued to produce detailed discussions that show how knowledge '[...] depends on being in a world that is inseparable from our bodies, our language, and our social history – in short from our *embodiment*' (ibid. 1991:149). The embodied enaction approach position developed by Varela et al acknowledged that there are many ways that the world is and can be experienced. Their alternative approach was born out of dissatisfaction with the basic assumptions of the cognitivist view: A world with particular properties which is accessed by the knower (a separate subjective 'we') through internal representation (ibid. 1991:8, 9). This representationalist position assumes a world with pre-given features which is mirrored by the internal representations of the mind (ibid. 1991:135, 136). Their alternative proposal emphasizes the importance of action – 'enaction'. They say that such systems do not operate by representation, they *enact* a world as a domain of distinctions that is inseparable from the structure embodied by the cognitive system (ibid. 1991:140). They argue that the world-cognition relationship is not one of representing or mirroring (agreeing with Rorty), but

> [..]more like a background – a setting of and field for all our experience, but one that cannot be found apart from our structure, behaviour, and cognition. For this reason we say the world tells us as much about ourselves as it does about the world (ibid. 1991:142).

For Varela et al, (1991:xvi) embodiment has a double sense. It encompasses both the body as a lived, experiential structure and the body as the context or milieu of cognitive mechanisms.

Cognition is not representation but embodied action which depends upon the kinds of experience that comes from having a body with various sensori-motor capacities that are in turn embedded in biological, psychological and cultural contexts. The world we cognize is not a pre-given world but one enacted through our history of *structural coupling* (Varela et al, 1991:201, 205, 206).[1] To the question, 'What is cognition?' Varela et al answer: 'Enaction: A History of Structural Coupling that Brings Forth a World' (ibid. 1991:206). Enactive Cognition as embodied *enaction* (ibid. 1991:172, 200) as approach or program, is intended to be a middle way between dualism (subject and object) and monism (the one or the other). It consists of two points: In the first instance perception consists in perceptually guided action and in the second instance cognitive structures emerge from the recurrent sensori-motor patterns that enable action to be perceptually guided (ibid. 1991:173, 202, 213).

1 The terminology 'structural coupling' was first used by Varela, 1988 (see Varela et al, 1992:151).

Varela et al's alternative proposal breaks the restrictive mode of cognitivism with its linear relationship between knower and that which is known in order to create space for the recognition of creative and dynamic, interactive, context-dependent know-how – '[…] the very essence of creative cognition' (Varela, et al, 1991:148). Grasping after an inner ground is the essence of ego-self and is in itself a moment in a larger pattern of grasping '[...] that includes our clinging to an outer ground in the form of the idea of a pre-given and independent world' (ibid. 1991:143). The search for an ultimate ground – the project of foundationalism – is not only limited to the subject and its basis in the ego-self, but it also includes our belief in a pre-given world (ibid. 1991:144) – a notion that they obviously do not share. On the contrary for these authors the world is not fixed or pre-given and tells us as much about ourselves as it does about itself (ibid. 1991:142, 144).

Two aspects of this position require some comment. The first is their rejection of the foundationalist programme which assumes the existence of a pre-given world and the second is the social constructivist alternative that they propose. Agreed, there is no pre-given world in the foundationalist sense, but it would be futile to argue that the way reality is structured does not somehow condition and constrain our knowledge of the world. That we come to know this world via creative human interaction can obviously not be denied, but human knowledge always bears the traits of *re*-cognition of that which is already there or at least potentially there. This *re*-cognition is not mirroring or re-presentation, but the dynamic interaction of a fully embodied person subjectively interacting with a contextually structured situation or relationship. The refreshing emphasis of these new and alternative views on the nature of knowledge is the recognition of the interactive subjective involvement of the knower and more specifically the role of embodiment in the wider and narrower sense in this process.

Yet another proponent of action in the process of meaning disclosure is Shanon, who argues against the representational-computational view of mind (RCVM) (1993:21ff) which assumes literal language as paradigmatic expression (1993:80). He demonstrates the consequences of the rejection of this view for the understanding of metaphor based on the so-called 'two-stage model'. Shanon (1993:89–91) takes issue with two basic assumptions of the three paradigmatic views of metaphor found in the literature (the substitution view, the comparison view and in the interaction view). The two assumptions are the 'fixedness assumption', which assumes metaphor is a relationship between pre-existing, well defined entities with fixed traits and the 'selectional assumption' which argues that '[...] manifest cognitive distinctions are defined through the selection of a subset from a given set of entities or values through fixing of values of such entities or of their weights' (ibid. 1993:89). In an earlier work Shanon claims (1992:674) that all three paradigmatic views of metaphor '[...] are couched in the assumption that the semantic constituents of metaphor are given and fixed and that metaphoric interpretation, even if novel, is achieved by means of a selection from them'. The 'fixedness assumption'[2] and the 'selectional assumption' are associated with two additional assumptions:

> [...](1) the primacy of atomistic constituents relative to higher order relations and *gestalts* (wholes), and
> (2) the elementary terms of cognition as skeletal structures, onto which meaning is projected by the process of interpretation.

2 Shanon (1992) discusses two assumptions of the representationalist- cognitivist view of mind that are based on the acceptance of a world consisting of set pre-given features.

Shanon surveys conceptual considerations and empirical data which argue against these four assumptions (1992:674) and argues that meaning is determined not by selecting among given features, but '[…] rather than selecting among given features, context creates features'. On empirical grounds – referring to speech act processing, neuropsychological considerations, developmental observations, artistic creation and cultural evolution – Shanon (1993:53) takes issue with the so called 'two-stage model' which argues that a standard meaning always precedes the non-standard or metaphorical meaning. He argues that metaphor generates new features and this happens through the creation of new contexts (1993:89). His argument is based on the 'weak definition' of metaphor which does not, like the 'strong view of metaphor' presuppose distinct entities that are already given (1993:61) but '[...] the figurativeness of categories' (that are) 'intrinsic'. This forms the basis of world views where every culture chooses a root-metaphor by which it views the world. These basic ontological components of the world are the very fundament of which the world is composed as e.g., the Kaballah acknowledges with respect to wisdom, justice, mercy. But, Shanon goes further when he states that '[…] metaphors are so useful because they mirror the workings of the mind' (1993:91; 1992:678). Central to this understanding of cognition is the fact that it is rooted in action (Shanon 1993:80). So instead of discussing cognition as representation and manipulation of the symbols of thought, this line of investigation sees cognition in terms of object manipulation, activities or skills and patterns of action (1992:670; 1993:81).

Shanon (1988:79) argues against the 'two-stage account of non-literal language' on four fronts. Conceptually the demarcation is unsatisfactory because both require contextual demarcation. Methodologically the distinction is not as parsimonious as was always thought. Procedurally '[...] the comprehension of non-literal expressions does not require the previous comprehension of the corresponding literal meaning' (1988:79). Developmentally the use of non-literal language seems to be an early phenomenon. Apparently, before becoming symbolic, language is embedded in the context of the child's action in the world. The latter notion is shared by Gibson's (1979) ecological psychology. This theory has as a cornerstone the notion that what people perceive '[...] are not atomistic sense data but, rather patterns of invariance'. The theory claims that '[...] these perceptions are defined in terms of higher-order, global patterns, not in terms of atomistic constituents' (Shanon, 1992:669) and thus metaphorical expressions, according to Shanon,

> [...] are open to being multiply experienced, but in themselves they transcend all of these ways of being experienced or read, and, inexhaustible, they still remain open to further associations and interpretations' (1992:678).

It is this open-ended contextual configuration of patterns that condition human action and interaction in the world which leads to the question whether metaphorical projections are indicators of underlying prototypical domains.

Metaphorical Projections: Constitutive of Domains?

In the earlier discussions concerning the nature of metaphor Johnson (1981:234) differentiates between a canonical and non-canonical level in the 'metaphoricity of thought'. In these discussions the notion of experiential structures, categorization and concepts already figure prominently, but the ontological issues which surface in the later discussions are only alluded to. In the discussion between Johnson (1993a) and Kennedy and Vervaeke (1993) the question concerning the constitutive role of metaphor is raised once more and in this discussion Shanon's so called 'fixedness assumption' surfaces, an assumption opening up a hornets nest of ontological issues.

Kennedy and Vervaeke challenge the claim that metaphorical projection can be constitutive of the conceptual structure of the target domain. They state that Johnson

> '[...]confuses mappings based on pre-existing features common to both source and target domains, with constitutive projections in which the source domain actually determines some of the conceptual structure of the target domain' (1993:416).

They argue that Johnson mistakenly conflates two claims:

> 1) that metaphors are based on pre-existing connections between the source and the target domains and
> 2) that the metaphors involve projections of structure from the source domain in order to structure the target domain (ibid. 1993:413, 414).

In response to Kennedy and Varela, Johnson argues for

> [...] the constitutive nature of conceptual metaphor which implies [...] that metaphor and other imaginative devices are matters of experience, understanding, conceptualization and reasoning rather than mere linguistic entities (1993a:414).

According to Johnson, 'Metaphors can be based on pre-existing structure, while at the same time partially structuring the target domain by mapping knowledge of the source domain on to the target' (1993a:414). If one recognizes that cognition cannot be understood without common sense and common sense in turn is none other than our bodily and social history then the '[...] inevitable conclusion is that knower and known, mind and world, stand in relation to each other through mutual specification or dependent co-origination' (Johnson 1987:150). I believe this conclusion to be central to any attempt at developing an account of the ontological conditions for metaphorical meaning. The critical question is how this 'mutual specification' or dependent co origination is anchored.

The Upshot?

From this brief overview it is apparent that a number of similar issues surface in the proposed anti-cognitivist alternatives: Are there fixed contours of experience and reality that dictate the itineraries of meaning which develop through metaphor? If there are such fixed contours how does one give an account of the creative and novel meanings that metaphor is able to create? Is 'embodiment' an adequate grounding for the possibilities of meaning variance and or meaning invariance? The puzzling factor that remains is the basis or grounding of the metaphorical meaning with the related notion of reality depiction and the cognitive claims of metaphor.

A common thread through the diverse non-objectivist alternatives to the representationalist-cognitivist view of knowledge is the emphasis on the dynamic character of knowledge formation and the *interactive* nature of the cognitive process. Deeply embedded in these discussions are questions related to the grounding of metaphorical meaning in either social or linguistic constructions, embodiment or some deeper ontological dimension.

The turn from the Cartesian objectivist, dualist understanding of cognition to one in which the embodied subjectivity of the actor is fully recognized has come about through a number of historical 'turns'. The Historical Turn in Philosophy of Science (Kuhn et al), the Linguistic Turn, the Sociological Turn and the Cognitive Turn have all contributed to the unmasking of the pretences of objectivism. Postmodernism with its critique of foundationalism and essentialism in turn has confronted us with profound and penetrating questions concerning the nature of reality, the nature of our knowledge of reality and the plurality of ways one can give an account of the nature of the world. This turn towards the subject and an openness towards the possibility of manifold interpretations have also created intellectual space for accounts in which religion legitimately figures.

Lakoff and Johnson, and Rohrer et al's introduction of embodiment and image schemata as the basis of cognition accommodates factors previously regarded as 'subjective' or non-epistemic in the account of the nature of cognition. The recognition of diverse but related domains of experience present in the embodied cognitive act, opens up new avenues of exploring the constitutive role of former 'non-epistemic factors' in cognition. Socio-cultural and ideological factors can now be accommodated in the act of cognition within scientific cognition. These developments coupled with the renewed interest in the sociology of knowledge and the cognitive-historical approach to scientific theorising (Brown, 1984; Collins, 1985; Hesse, 1983 & 1988b; Nersessian, 1987; 1991a; Tweney, 1989b) open avenues to develop a different and better understanding of the relationship between so called 'non-epistemic' factors and the 'epistemic' enterprise of science. The new insights into the nature of cognition have opened the possibility of acknowledging the nature and presence of religious influence intrinsic to and constitutive of the cognitive act. The consequences of these developments for an account of the role of religion and faith in the scientific enterprise will be explored in a later chapter. I will argue that in such a view the understanding of metaphor, analogy and related phenomena require a stronger view of cognition based on an understanding of the nature of 'embodiment' that challenges both the standard objectivist views of reality and knowledge and the anti-objectivist alternatives. It will also need to take issue with the position of the representational and objectivist view of knowledge which assumes a pre-given world populated with 'natural kinds' endowed with fixed features, that are somehow 'represented' by the cognitive agent. Obviously not all these matters can be addressed here. Some I have already addressed (Botha, 1986b). A significant aspect of this discussion now calls for our attention: the grounding of metaphorical meaning and its relationship to domain theory and categorization.

II Solid Ground? The Weight of Grounding Hypotheses

In this chapter I attempt to explore some of the alternative views concerning the grounding of metaphorical meaning. Currently the embodiment thesis is one of the strongest contenders for the grounding of metaphorical meaning. I will attempt to articulate the contours of an alternative theory of embodiment which grounds metaphorical meaning and meaning change in an ontological and anthropological framework. This is a framework that recognizes the presence and active functioning of multi-ordinal domains and (ordered) structures for reality. This proposed view of embodiment purports to provide a basis for the phenomenon of cross domain mapping, blends and the prevalence of multi-ordinal concepts in the disciplines. This will form a prelude to the understanding of the constitutive role of fiduciary beliefs in the act of theoretical cognition, which will be dealt with later.

I will show that this view of embodiment and cognition provides an alternative for understanding the relationship between faith and science by facilitating a revised view of the way 'religious' beliefs operate in theorizing. This interpretation of embodiment accommodates a specific understanding of the hermeneutical role that metaphor and analogy play in the process of imaginative discovery in the world and in the natural and the social sciences. Basic to this proposed notion of embodiment is an ontology which recognizes the structural stratification and categorization of both human experience, the human body and reality. These categorizations condition and constrain analogical and metaphorical meaning transfer, cross domain mappings and blends in cognition and in language. Before embarking on this alternative, we need to have a closer look at the problem of the grounding of metaphorical meaning and the role domains and categories play in this process.

The 'Grounding Hypothesis'

Searle (1988:141-159), Ricœur (1977b:159), Johnson (1981:200); and Lakoff and Johnson (1999), all argue for some or other form of recognition of non-propositional and pre-conceptual structures as basis for (metaphorical) meaning. Their arguments differ on fundamental points: their adherence to or rejection of the causal theory of reference, views concerning the existence of natural kinds and the standard logic-and-literal language view with its corollary 'objectivist thesis' concerning the nature of knowledge. These authors emphasize a deeper or more fundamental analysis of the nature of analogical reasoning and meaning transformation across diverse contextual fields, the need to ground metaphorical meaning in embodied or concrete experiential *gestalt*s and to accommodate the place and role of so called 'non-epistemic' factors in theorizing.

Recent developments in Cognitive Semantics (Johnson, 1987; Lakoff & Johnson, 1988:119–154 and Johnson, 1993b:61) and cognitive semiotics address the problem of how metaphorical meaning is possible through discussions of the 'grounding' of metaphorical meaning. The problem of the ontological grounding of cognition surfaces in Lakoff and Johnson's rejection of Rorty's relativism, and in the postulate of an experiential basis of meaning and the dynamic interactive character of the nature of knowledge formation. Lakoff and Johnson's (1999) work in this area represents one of the many attempts at 'grounding' meaning in embodied experience. They propose to develop a responsible philosophy which is grounded empirically not grounded in *a priori* assumptions. How successful this project is, is open to discussion and requires closer exploration.

Where metaphors allow us to understand one domain of experience in terms of another (Lakoff & Johnson, 1980:117), it is generally argued or assumed that metaphorical understanding is grounded in non-metaphorical understanding (Lakoff & Turner, 1989:113), an assumption characteristic of most reductionist theories of metaphor. The question raised here is to what extent Lakoff, Johnson and Turner's move away from grounding metaphorical meaning in literal language to autonomous concepts grounded in patterns of bodily and social experience, actually solves the problem without succumbing to either subjectivism or materialism. I argue that in order to escape the trap of subjectivism the move to ground metaphorical meaning in bodily experience requires a further recognition of the ontological and anthropological stratification which conditions and constrains meaning and meaning variance. Philosophical insights from the philosophy of the Dutch legal philosopher, Herman Dooyeweerd's so-called *Amsterdam School*, will be utilised in this analysis. Although the terminology differs Dooyeweerd's views on analogies bear some resemblances but also fundamental differences to Lakoff and Johnson's 'conceptual metaphor' theory.

The Conundrum of 'New' Domains

One of the challenges to reductionist theories of metaphor is the fact that it does not provide a convincing explanation for the fact that through metaphor new domains of meaning are opened up. The traditional position concerning the creation of new metaphorical meaning does raise some significant ontological questions. One basic question is whether the metaphorical projection between source and target domains actually do '[...] give rise to a new structure' (1993a:417). The issue here is how pre-conceptual structures (domains) can conceptually give rise to new 'domains', especially if the structure of the body constrains and determines the direction of metaphorical mappings. If one argues that these cross-domain mappings are none other than 'mental spaces' one can avoid answering this thorny issue. 'Mental spaces' allow for manifold proliferation of possible blends, whereas an approach which argues for the fixedness of body traits of necessity needs to come up with an exhaustive list. The underlying issue here is whether the meaning of the metaphorical projections are based on pre-existing features of the body and the world or are constituted in the process of metaphorical meaning transfer, or perhaps both. On the one hand the issue relates to the much older Richards-Black distinction between *epiphors* and *diaphors* – an issue that I have dealt with extensively in my article on the 'Framework for a Taxonomy of Metaphor' (1988a) and also tangentially explored in the chapter on 'Metaphorical Models and Scientific Realism'. On the other hand it relates to a more recent discussion concerning the nature of representation in cognition. I will tie the second issue to the more recent embodiment debates and the alternative proposal that I intend to explore. Lakoff and Johnson's 'conceptual metaphor theory' will function as foil for the proposal.

In the earlier work of Lakoff and Johnson (1980:56) the problem of grounding the conceptual system was briefly articulated as follows: 'Are there any concepts at all that are understood directly, without metaphor? If not, how can we understand anything at all?' Concepts that are candidates to be understood directly are plants, departures, fire, sleep, locations, seeing, etc. Lakoff and Johnson claim that a difference exists between their grounding hypothesis and the literal meaning theory. Their 'grounding hypothesis' which is about concepts and not about language, indicates that only *some* concepts are 'semantically autonomous', yet *most* concepts are not semantically autonomous (Lakoff & Turner, 1989:119). In their position the term *literal* has been relegated to be used as handy term either for a source domain of a metaphor or to contrast with such terms as 'ironic,' 'understated,' 'arrived at by principles of the conversation', etc (ibid.,1989:119).

Lakoff and Turner argue that the literal meaning theory is not about concepts but about ordinary, conventional language which is seen to be semantically autonomous. Their grounding hypothesis deals with concepts as embedded in human experience. They do not deny that there are semantically autonomous concepts but claim that whatever such concepts, they '[...] are grounded in our patterns of bodily and social experience' (ibid. 1989:119). To these issues and the issue concerning the grounding of orientational, basic, ontological and structural metaphors they answer that they are grounded '[...] by virtue of systematic correlates within our experience' (Lakoff & Johnson, 1980:58, 61). In order for metaphors to allow understanding one domain in terms of another, there must be some '[...] grounding, some concepts that are not completely understood via metaphor to serve as source domains' (Lakoff & Turner, 1989:135).

Lakoff and Johnson (1980:59) emphasise the need to differentiate between experience and the way it is conceptualised and argue that '[...] there are natural dimensions of experience and that concepts can be analysed along these dimensions in more than one way' (1980:76). They argue that understanding takes place in terms of entire domains of experience and not in terms of isolated concepts. They claim that domains of experience are conceptualised as experiential *gestalts* – structured wholes – that represent a coherent organisation of experiences in terms of what is experienced as 'natural kinds of experience' (Lakoff & Johnson, 1980:117). Their analysis of a host of empirical examples leads them to their *conceptual metaphor* theory which assumes the existence of these experiential domains. In *Philosophy in the Flesh* (1999) where they inter alia deal with the grounding of metaphorical meaning, they propose an empirically grounded (responsible) philosophy which is not grounded in *a priori* assumptions. The target of their attack is the dominant objectivist treatment of language, meaning, understanding and reasoning which does not take into consideration the bodily experience and figurative process of ordering.

In *conceptual metaphor theory* metaphors are analyzed as stable and systematic relationships between two conceptual domains, whereas in *blending theory* the basic unit of cognitive organization is not a domain but a 'mental space' (Fauconnier, 1994:16; Oakley, Grady and Coulson, 1999). Lakoff and Johnson (1980:116) are primarily concerned with how people understand their experiences. They state that language provides general principles of understanding which involve whole systems of concepts, rather than individual words or individual concepts. A standard view of definition, they claim, seeks to be 'objective' and it assumes that experiences and objects have inherent properties and that human beings understand them solely in terms of these properties. Against this view, they argue that an event or state of affairs, e.g. 'love', is comprehended only partly in terms of such inherent properties. For the most part our comprehension of love is metaphorical and we understand it primarily in terms of concepts for other natural kinds of experience, e.g. 'love is a journey', 'love is madness', etc. This leads them to the interesting distinction between *inherent* and *interactional* properties of concepts (1980:119). These interactive qualities are ascribed to objects, events and activities through human interaction with them, such as perception, motor activity, function, etc. An 'objectivist' or 'set theoretical' concept of categorization they claim, does not give sufficient account of the fact that interactional properties are prominent among the kinds of properties that count in determining sufficient family resemblance (1980:123). They argue that individual concepts are not defined in an isolated fashion, but rather in terms of their roles in natural kinds of experiences. Concepts are not defined solely in terms of inherent properties; instead, they are defined primarily in terms of interactional properties (1980:125). Presumably this is the state of affairs that Hesse is also referring to in her account of the metaphorical nature of all language. Conceptual metaphor theory has proven to be most useful and fertile for the understanding of metaphorical meaning creation. It is, however, exactly the architecture of these domains (Brandt, 2000) at stake in the processes of cross-domain mapping and meaning transfer through metaphor, that will be explored in this chapter.

The nature of 'domains' require closer exploration, in order to clarify the nature of domains and to delineate the relationship between conceptual metaphors, the experiential *gestalts* they are based upon and the pre-theoretical and pre-conceptual nomic conditions which condition and constrain domains.

Conceptual metaphor theory claims that '[...] the magnificent tool of conceptual metaphor' (Lakoff & Johnson, 1999:233) '[...] is one of the greatest of our intellectual gifts' (ibid, 1999:45, 129). Metaphors are pervasive both in thought and in language, and in human subjective experience in general. A large system of primary metaphors is acquired automatically and unconsciously by functioning in everyday life. 'Because of the way neural connections are formed,' Lakoff and Johnson (1999:47) state '[...] humans all naturally think using hundreds of primary metaphors'. They also endorse the view which is contrary to long-standing opinion, that primary metaphor is not the result of a conscious multistage process of interpretation, but is a '[...] matter of immediate conceptual mappings via neural connections' (ibid. 1999:57). They (1999:46, 47) develop an 'integrated theory of primary metaphor' which includes four parts: Johnson's theory of conflation in the course of learning; Grady's theory of primary metaphor; Narayanan's neural theory of metaphor and Fauconnier and Turner's theory of conceptual blending.

Lakoff and Johnson differentiate a set of metaphors they call 'ontological metaphors'. Ontological metaphors are the type of metaphor where abstract notions are thought of as concrete entities or substances. I would use the term 'ontological' in a different sense. I prefer 'ontic' for this level. Ontological is a theory or philosophy about that which is ontic. Lakoff and Johnson's emphasis on the bodily nature of cognition also takes into account the multifaceted nature of the subjective experience of the world which forms the basis of analogical and metaphorical meaning change and meaning transfer. Their notion of embodiment and conceptual metaphor is based on an important assumption that '[...] our corporeality is part of the corporeality of the world [...]' (ibid. 1999:565). The evidence they provide is based on extensive empirical analysis of 'conceptual metaphor' and conceptual domains. When Lakoff and Johnson (1999:462) discuss the grounding of conceptual metaphor they argue that second-generation cognitive science locates meaning in the body and the unconscious conceptual system, and that meaning arises in the body and brain through our interactions with the environment and with other people (ibid. 1999:463).

Lakoff and Johnson tie this innate human propensity to the embodied spirituality of human beings. This in turn comes to expression in what they call empathic imaginative projection – a form of 'transcendence, [...] a form of being in the other [...]' (ibid. 1999:565) and in 'ecological spirituality' ibid. 1999:566) in the relationship to the physical world. 'Embodied spirituality' entails that in all acts of imaginative, empathic interaction with the world around us an element of what has traditionally been called 'the spiritual', is present (ibid.1999:565). The fact that the analysis of meaning disclosure of metaphor leads to the recognition of the presence and role of embodied spirituality is an issue that will be dealt with elsewhere.

Much of what Lakoff and Johnson argue opens up new avenues to understand the bodily basis of metaphorical meaning. The introduction of 'conceptual metaphor' theory provides significant access to the notion of conceptual domains and the grounding of metaphorical meaning in such domains. But are these domains only corporeal? How is the corporeality of humankind and that of the world related? These and other questions will be pursued in the following sections.

An Ontological Framework Conditioning the 'Itineraries of Meaning'

My provisional agreement with Lakoff, Johnson and others about the 'conceptual' nature of metaphor rests on an important caveat, viz. that this bodily based conceptual structure which lies at the basis of linguistic articulations of conceptual metaphor, is grounded in a deeper ontic structure for the world and of human experience. It is the 'metaphorical' (actually 'analogical') ontological structure of this grounding that needs to be 'fleshed out' in a more encompassing architecture of domains. In such an architecture the 'multi-ordinality' or multi-facetedness of reality, facts, things, human and social relationships, events, human action, human experience and cognition in diverse domains of experience, need further adumbration. Such a framework must be able to ground the notion of 'conceptual metaphor' and meaning in the ineradicably relational nature of both human beings and reality and in the relationship of human beings in and to reality. In order to escape the pitfall of subjectivism such a framework needs to accommodate the fact that concepts, conceptual domains and metaphors are constrained and conditioned by a deeper, ontic framework which conditions the 'itineraries of meaning' (Ricœur, 1980), which guide and structure the acts of knowing. These 'itineraries' are not only constitutive of the human mind, but also of the nature of the world. There is indeed a correlation between the corporeality of human beings and the corporeality of the world. In order to escape the trap of materialism, such a framework needs to recognize the bodily basis of human experience, conceptualisation and language without reducing this to its neural basis. This is not only an argument for anti-reductionism. It is also an argument for the recognition of the deeply spiritual mooring of human existence and cognition. This spiritual mooring reveals itself in the open-ended process of metaphorical meaning disclosure and the unique constraining, limiting and guiding role of the multiplicity of diverse laws conditioning meaning.

All human experience is conditioned and constrained by pre-conceptual and pre-linguistic basic level ontic categories which determine the boundaries between analogical and non-analogical or 'original' meaning (Dooyeweerd, 1954; Hart, 1984). The mere possibility of speaking about category mistakes rests on an assumption that the world we deal with is in some way categorized. Whether these categories are socially constructed and the result of culture and convention or whether they are representative of deeper and more fundamental universal structures of the world, both views imply that categorization is basic to logic and language. This means that so called category mistakes can only be identified on the basis of the acceptance of some notion of a stratified reality. The grounding of these strata in social convention often becomes apparent in the examples of language use in e.g. a Thesaurus. But basic to the conventionalized categorizations of language are underlying ontic constraints. Lakoff and Johnson have provided us with many examples from everyday life. They (1980:19, 117, 118; 1982b:193ff) argue that metaphors are conceptualized as experientially basic *gestalts*; are products of human nature and form multi-dimensional structural wholes. Some may be universal, while others will vary from culture to culture. The types of experience analyzed by Lakoff and Johnson include inter alia, spatial, physiological, psychological, mental, social, political, economic and religious experiences. They suggest that the concepts which appear in metaphorical definitions are those that correspond to these natural kinds of experience. Johnson (1987:30) elaborated these insights by arguing that meaning, understanding and rationality are constrained by non-propositional 'image-schematic' structures such as e.g. temporal and spatial orientation which arise from our bodily experience and which are metaphorically projected and extended in order to understand reality.

It is the origin, nature and functioning of these structures, that constrains metaphorical and analogical understanding that Johnson explores in *The Body in the Mind* (1987) and in 'Some constraints on embodied analogical understanding' (Johnson, 1988:25). The account he develops on the basis of a variety of examples show that there are common structures of imagination that emerge from our constant bodily functioning (1988:33). In his *The Body in the Mind* (1987:42–48) Johnson explores the way in which patterns of typical experiences of force work their way into the system of meaning and into the structure of expression and communication. The main evidence for the efficacy of these image-schematic *gestalt* structures, he derives from the way such a structure constrains and limits meaning as well as from patterns of inference in human reasoning.

Perhaps the most salient point where the need for the delineation of ontic boundaries becomes acute is the challenge to identify ideologies. Any attempt at diagnosing metaphorical hypertrophy (internal ideological derailment) in science necessarily has to appeal to some philosophical interpretation of the limits set by these ontic boundaries. In the next section some empirical examples from everyday experience and from social science illustrate the contextual qualification and differentiation of an image-schematic structure. The vagaries of the metaphor and multivalent analogical concept 'force' in the natural sciences is also a case in point.

Discerning Recurring Regulated Patterns

One's access to these ontic constraints and conditions are by means of human embodied experience and through conceptual, linguistic or imaginative articulations. The constraints are not created in these processes. Hart (1984) illustrates this point with a good example. If one were to fly over a major city and observe the obviously regulated patterns of traffic flowing beneath you, you would infer that there are traffic rules, and moreover, that these rules have been made by some competent body. When we recognise the fact that meaning is discerned based on recurring regulated patterns, we are dealing with an analogous state of affairs. Discerning regularity and pattern and *gestalts* which constitute the basis of meaning does not imply that the meaning was created by the discerner.

Lakoff and Johnson's anchoring of meaning in bodily existence falls short of actually recognising that all realms of reality are permeated by and pregnant with meaning which the knower in community with others *opens up and discovers* via human interactive experience. It is exactly this dynamic '[...] intrinsic restlessness and relational insufficiency of reality' (Hart, 1984:166), which human action and cognition participate in and which points to the 'expressive' and 'referential' character of all of reality where '[...], meanings refuse to stand still' (Van Hoozer, 1998:127). Both conceptual metaphor and the states of affairs on which it is based in reality have this deferral of meaning built into them. Having said this, it does not exempt us from the responsibility to track down the mechanisms by means of which we approximate such meaning. It is to the contours of this notion of 'domains' that we now turn.

Domains: Structured Mush?

Lakoff (1987:268) recognises that a satisfactory account of meaning and meaningful understanding rests on the recognition of the structured nature of embodied, human experience. Experientialism, Lakoff claims (1987:267), rests on the bodily pre-conceptual form from which it arises and this is not 'unstructured mush' (ibid. 1987:267). Lakoff and Johnson (1980:117, 118) argue that there are three natural kinds of experience: of the *body*, of the *physical environment*, and of the *culture*. They regard them as 'natural' because they are products of human nature and form multidimensional structural wholes. (ibid. 1980:116). There is a relatively small number of conceptual metaphors drawing on domains of bodily experience which structure abstract human concepts (Johnson, 1989:115). Johnson claims that our knowledge is embodied in a deeper and more profound sense than mere know-how. Our conceptual system '[...] is grounded in and structured by various recurring patterns of our perceptual interactions, bodily *orientations*, movements and manipulations of objects' (Johnson, 1993a:414). He states that this '[...] corporeal semantics, a semantics of embodied understanding' (ibid. 1993a:422) actually is constitutive of our cognitive activity and of our concepts. In their view human embodied nature and cognition are patterned and orderly, and this order reflects in the concepts formed in and about the world.

In the early work by Lakoff and Johnson (1980:17–19) *Metaphors We Live By*, the authors develop a thesis that '[…]a 'basic domain of experience', '[...] is a structured whole within our experience that is conceptualized as '[...] an experiential *gestalt*'. They represent coherent organizations of our experiences in terms of natural dimensions. Domains of experience that are organized as *gestalts* in terms of such natural dimensions seem to us to be natural kinds of experience (Lakoff and Johnson, 1980:117,118; 1981b:193ff).

Some of these experiential *gestalts* may be universal, while others will vary from culture to culture. The types of experience they analyze include inter alia, spatial, physiological, psychological, mental, social, political, economic and religious experiences. They suggest that the concepts which appear in metaphorical definitions are those that correspond to these natural kinds of experience. They say: 'The kind of conceptual system we have is a product of the kind of beings we are and the way we interact with our physical and our cultural environments' (Lakoff and Johnson, 1980:119). They reject the idea of the grounding of the conceptual system within the framework of a so-called 'building block' theory in which all meaningful utterances are constructed from certain unanalyzable semantic units. Instead they identify emergent categories and concepts that are best understood as experiential *gestalts*, which, though decomposable into other elements are yet basic and irreducible in terms of grounding our conceptual system (1981a:314–5). It is the nature of the pre-conceptual reality that needs further reflection.

Metaphors always rest on some sort of analogy that relates to differences and/in similarities, within domains or between different domains. Metaphors are not only lingual articulations of purported interactive meanings between two different semantic domains or between differences and similarities within one domain, but are also typical of thought processes about the world. They express more than merely lingual states of affairs, but rely on ontic categories, (natural) kinds and image schemata, and in turn 'create' (Fauconnier, 1994) domains or 'mental spaces'. Image schemata are pre-conceptual experiences that occur constantly in everyday life, e.g. orientations and relations (up-down, part-whole, centre-periphery, etc. (Lakoff, 1987:267).

Domains and Mental Spaces Not Only Creations of the Human Mind

No doubt the human mind is actively involved and creative in the process of uncovering and opening up existing and new domains. The recognition of this creative act need not commit one to a position in which the existence or creation of these domains are ascribed primarily to the outcomes of neural mechanisms. That our recognition and knowledge of such domains are based on effective neural processes requires little argument, but this does not imply the neural origin of such domains. The tendency toward transcendence mentioned above is an indication that even in our most creative and imaginative moments we succeed only in gaining limited insight into a part of the deep and unfathomable complexity of meaning which precedes our cognitive or linguistic grasp and which seems always to be unfolding more layers and depths of meaning. A reading, which ultimately attributes 'domains' and 'mental spaces' to creations of the human mind does not satisfy, because according to this view, meaning is merely the result of human physical and mental processes. All these processes play a significant role in the uncovering of meaning and change, but fail to provide an adequate and convincing grounding of meaning. Meaning and metaphorical meaning can not be a bootstrap operation in which all resources of meaning are attributed to the subject or located in the material, neural basis. On the contrary, meaning is presupposed when human actors interact with the world. It is to the possible structure of the domains in which this presupposed meaning comes to expression that we now turn.

Other Domain Theories

Eve Sweetser

The early relatively simple definition of metaphor as the understanding of one domain of experience in terms of another domain of experience raised the fundamental question: What are 'domains of experience'? (Lakoff & Johnson, 1980:117; Brandt, 2000:12). The notion of a *domain* is found in the earliest literature on metaphor, e.g. in terms like *source domain* and *target domain* but is not explicitly defined. Brandt (2000:11–51) shows that a few authors (Lakoff & Johnson, Turner, Sweetser) have done work in this area, but argues that no adequate account of domains has yet been developed. He then proceeds to develop an 'architecture of semantic domains as grounding hypothesis in cognitive semiotics' (see Brandt, 2000). This proposal is preceded by Eve Sweetser's (1990:19) analysis of semantic change and polysemy patterns in which she introduces at least four systematic metaphorical connections between domains: social, physical, mental and speech acts. Sweetser states the following:

> Metaphor operates between domains. It operates so pervasively that speakers find an inter-domain connection between knowledge and vision, or between time and space, to be as natural as the intra-domain connection between *finger* and *hand* or between *man* and *woman* (1990:19)

Per Aage Brandt's proposal for an architecture of semantic domains on the other hand, provides an elaborate analysis of the variety of domain addresses that provide accommodation to metaphor concepts.

Brandt presents a 'geography of the life-world' and provides an 'architecture of semantic domains as a grounding hypothesis in cognitive semiotics' (see Brandt, 2000). He claims that we are embodied according to different basic domains of reality, calls for a return to the inaugural studies of metaphor and shows that 'metaphor concepts are superordinate semantic indicators of domain addresses' (Brandt, 2000:48). This approach chooses gesture, action, exchange and discourse as the four main categorisations of domains. In his 'first life-world map' he distinguishes a total of four basic *gesture-based* semantic domains: the physical, the social, the mental, and the speech-act domains (D1, D2, D3 and D4) and three *action-based* satellite or practical domains: work, love and worship (D5, D6 and D7). To this he adds a second and third satellite generation of domains that are *exchanged-based* (D8-D10): jurisdiction, economic exchanges, aesthetic evaluations in cultural life, and three fundamental *genres of discourse* (D11-D13): argumentative, narrative and descriptive (Brandt, 2000:45). To what extent such a categorisation is productive of actual insight into the process of meaning creation obviously depends on the type of problems it is able to solve or the number of approaches it can accommodate. I avoid a critical discussion of these two approaches in order to give attention to an alternative which accommodates the idea of ontic conditions basic to the distinction of domains and meaning.

We now have a number of arguments that allude to the existence of some ontological, experiential and/or embodied grounding of metaphorical meaning and meaning change. These positions have in common that they deny the existence of a '[...] rock-bottom core of literal concepts to which all meaning or conceptual structure can be reduced', yet do suggest some basic experiential grounding (Johnson, 1993a: 421). It is about the nature of the embodied experiential grounding that views diverge. All three sets of proposals above have merit and share one common denominator, that is, their approach to the grounding of metaphorical meaning from the angle of cognitive linguistics and cognitive semantics (Brandt would prefer to call it cognitive semiotics). Brandt's proposals have a Continental philosophical flair which reminds strongly of Husserlian phenomenological (life-world) roots, but his analyses are very much in the style of cognitive linguistics with its close attention to sentence structure and meanings of single concepts. Brandt has left 'domain theory' behind and wants to work with 'mental spaces' (along the lines of Fauconnier), presumably grounding the nature of meaning in the mental capacities of human beings. Brandt's (2000) proposal and methodology of an architecture of semantic domains recognises the bodily basis of semantic domains and multi-modal *gestalts* (Brandt, 2000:19), and develops a taxonomy of domains in the proposed 'first life-world map' (Brandt, 2000:13). His appeal to the 'life world' is an important step in the right direction as is the proposal of Lakoff and Johnson which grounds metaphorical meaning in experiential *gestalts* and human embodiment. Here too it is apparent that meaning comes to expression in a variety of speech acts, facts, events, things, societal relationships etc., and that all incorporate some form of metaphorical language, and that meaning resides in domains of reality. Brandt's analysis of domains presupposes the anthropological basis. Gesture, action, exchange and discourse are human functions. This remains the case when he chooses to work with the idea of 'mental spaces'. This anthropocentric emphasis is also true of Lakoff and Johnson's grounding of metaphorical meaning. The question remains: What is the role of categorization and domains in the process of grounding of metaphorical meaning. This issue is addressed by Turner (1988), Johnson (1988), Lakoff (1987) and Hesse (1988a).

In Rosch's experimental work on and development of a theory of categorization she argues for prototype members of categories that act as *cognitive reference points* (1975a:532–547). These are similar to Wertheimer's (1938) suggestion that there are certain 'ideal types' which act as the anchoring points for perception or conception (1975a:532,544. Cf. also Rosch et al, 1975b). It is the nature and role of these anchoring or mooring points that are of interest here.

Turner (1988) argues that principles of categorization lie at the root of metaphor and analogy and reflect the deeper ontic categorial structures. In this respect domain theories and theories of categorization are closely related. Turner (1988:3) notes: 'Our categories cut up our world into clusters, so we can deal with clusters instead of with elements. The fitness of our cognition will depend upon the fitness of these conceptual cuts'. What is regarded as an analogy will be closely related to and to a high degree determined by our category structures.

Johnson deals with some of the constraints on embodied analogical understanding and focuses upon the question where the structural isomorphism between source and target domains comes from and how it comes to expression in analogical projections. He says when analogy is treated as

> [...] principally a conceptual and propositional structure that can be sententially represented [...] it becomes virtually impossible to make the case that an analogy is irreducibly constitutive of our experience and understanding [...] (Johnson 1988:25).

He argues for the need for a theory of embodied analogical imagination and then develops a thesis arguing that '[...] analogical and metaphorical projection is pervasive in human understanding at a level of meaning and reasoning below that of propositional relations' (Johnson, 1988:26). He says our theorizing is inadequate because it does not give sufficient attention to the pre-conceptual and non-propositional levels of cognition. For this reason no propositional analysis alone is ever completely adequate (Johnson, 1988:28).

Johnson shows that Searle, Davidson and Ricœur agree that '[...] we must explore levels of metaphorical projection that go beyond anything propositionally representable'. Johnson's research has led him to what he calls 'schematic structures of understanding' (1988:29), which he regards as '[...] an abstract structure of an image [...]' or '[...] as the recurring structure or pattern of our imaginative process of forming an image'. Examples of such image schemata are PATH, BALANCE, FORCE, CYCLES, DEGREE etc. Johnson (1988:33) says analogical understanding is indispensable and '[...] reaches down more deeply into the embodied schematic imaginative structuring of our experience, where form, pattern and significance first merge for us'. These irreducible analogies are constitutive of our experience because they are partially constitutive of our understanding [...]' (Johnson, 1988:39). Turner (1988:5) speaks of 'indispensable analogies' which makes a bid to establish or influence category structures'. He argues: 'A powerful analogy can re-structure, disturb, influence and change our category structures' (ibid., 1988:5).

Elsewhere (1987:104,105) Johnson says that understanding is the way we have those experiences in the first place (not the after-the-fact reflections on prior experiences). He says: 'It is the way our world presents itself to us'. This is the result of metaphorical structures that blend all of the influences (bodily, perceptual, cultural, linguistic, historical, economic, etc). 'Metaphor reaches down below the level of propositions into this massive embodied dimension of our being' (Johnson, 1987:105)

Two important themes follow from the above analysis: The first is the question to what extent the way the world presents itself to us is the *result* of bodily, perceptual, linguistic, historical or economic influences and to what extent our understanding is actually *conditioned* by these life-constitutive categories and domains – an issue that will be dealt with later. The pre-conceptual, non-propositional categories are constitutive of this lived embodied understanding and are the mooring of the multivalent radial structures of reality. The second theme relates to the question concerning the nature of the analogical relationship between categories and domains. Lakoff and Johnson's notion of prototypical radial structures will prove to provide a significant access to the exploration of this problem.

Lakoff and Johnson's emphasis on embodiment as the basis of meaning is a choice for an anthropocentric grounding. They claim that it is mistaken to think that the spatial relations our conceptual system relies upon and which we take for granted, are just '[...] objectively given features of the external world' (Lakoff & Johnson, 1999:575). They argue that we make the best of what our brain and our visual system offer us of this world. In this quote, as in other sections of the 1999 major work of Lakoff and Johnson, embodied realism locates metaphorical meaning in the body, the brain and the neurological functioning of the human brain. Their view raises important questions:

It seems to lack recognition of the stratified nature of non-human reality, which correlates with the stratification of human embodied experience. They locate and explain metaphorical polyvalence based on embodiment without recognizing the pre-given ontic basis for this multivalence which is also characteristic of the world. The stratified 'corporeality' of the world requires some deeper ontological grounding than grounding merely in human mental capacities. There is also another important dimension of metaphorical meaning that is not adequately addressed in these approaches: A certain type of conceptual metaphor expresses the analogical radial structure of meaning harboured in diverse aspects of reality. There is a need for a theory that explains the fact that this type of conceptual metaphor embodies what Clouser (1991:215) calls 'aspectual universality'. This is the phenomenon that all domains seem to echo or reflect all other domains.

The Concept of Radial Structure

An important clue to both the problem of conceptual and metaphorical multivalence and polysemy on the one hand, and the ontological bases for the principle of aspectual universality on the other, is to be found in the idea of the 'radial structure' of categories. The concept of radial structure as I propose to use it, differs from the way it is used by Lakoff and Johnson. Radial structure should not be seen as predominantly a conceptual matter, but primarily an ontological (actually 'ontic') matter. Reality, both in its entitiary and aspectual existence, reflects the diverse nuances of the coherence of meaning. It is the ontological assumptions of such a conception of *domain* grounding of knowledge[1] and metaphor that requires further fleshing out. I propose to deal with this in the next section. In this section I have broken with the double-language thesis and its conventional grounding of meaning in the literal. I have also broken with an understanding of embodied realism which denies that we live and experience a world 'given' to us. There is no 'earthly world'-in-itself, but only one that stands in a

1 The idea of a *domain* was mentioned by Lakoff in 1987.

'concentric relationship' (Dooyeweerd, 1955:549) to humankind.[2] But this view cannot imply that all facets of reality and all human experience of reality can ultimately be grounded in neural processes or human embodiment.

In a radial category there is one central case and the others are extensions of the central one (Lakoff and Johnson, 1999:224, 225). The example which Lakoff and Johnson use to illustrate this is 'harm'. The central kind of harm is physical, but there is also emotional, financial and social harm which form metaphorical extensions of the central case. They may represent metaphorical extensions of the prototype (as in 'harm'), or alternatively, categories that are missing one or more features associated with the prototype (as in 'mother'). Lakoff & Johnson (1999:224) claim that our very concept of causation is multivalent: It consists of the entire radial structure, with human agency at the centre and many extensions. The word 'causation', they say '[...] designates a human category, a radial category of extraordinary complexity'. We have a central prototypical case of what 'causation' is in our physical experience of for example being punched in the arm, but, Lakoff & Johnson argue,

2 This calls forth a number of philosophical issues which cannot be dealt with in this chapter. Within the tradition of Reformational Christian philosophy there has been a significant difference between the positions of Stoker (1967) and Dooyeweerd (1957:74, 75) on this matter: The former argues that all realms (inanimate, vegetative, animate and human) have their own intrinsic meaning and relationship to God. Dooyeweerd on the other hand, argues for an anthropocentric, (con-) centring of reality in human embodiment and relations. This stance in turn is grounded in the transcendent and transcendental *a priori* structure of reality which conditions human experience and makes human experience and knowledge possible (Dooyeweerd, 1955:548).

> the question is, however, problematic just about everywhere else, because we are moving away from the central prototypical case of causation to other very different senses with different logics and different criteria for determining what is true(1999: 233).

Thus causality differentiates according to a great variety of contexts or aspects of human experience and reality. The abovementioned differentiation according to a variety of contexts raises the question concerning differentiation between the (centrally prototypical) meaning of causality and the differentiated analogical meanings of causality which are found within the radius of the domain inhabited by the prototypical case. How is this determined? Some ontological distinctions found in the philosophy of Herman Dooyeweerd are helpful in this respect. They will be pursued in the following chapter. But before we introduce his distinctions some brief notes on cross-domain mappings and radial categories as developed by Lakoff and Johnson et al.

Cross-domain Mappings and Radial Categories

In the vocabulary of Cognitive Science the term *domain* is used in a variety of ways. This includes domains of conceptual metaphors, linguistic domains, semantic domains, domains of experience, etc. 'Cross domain conceptual mapping' (Lakoff & Johnson, 1999:71) is described as a 'cognitive mechanism' which is based on the existence of conceptual metaphor, an embodied system of 'basic-level concepts' that '[...] have evolved to 'fit' the ways our bodies, over the course of evolution, have been coupled to our environment' (Lakoff & Johnson, 1999:91). Each metaphorical idea which is binary and not unitary harbours a cross domain mapping – '[...] it has both a source and a target that is at least partly structured by that source' (Lakoff & Johnson, 1999:255). Fundamental to the domains are categories and prototypes. A *prototype* is the most central, or typical instance of a category. Whereas 'radial categories' are extensions of the prototype, they are less 'typical' and may differ from the prototype in one or more features.

Lakoff and Johnson (1980:117) describe the nature of a domain of experience as 'natural kinds of human experience'. They (1980:117, 118) argue that there are three natural kinds of experience viz. 'experiences of the body, of the physical environment, and of the culture'. They constitute the source domains upon which metaphors draw. Understanding takes place in terms of entire domains of experience and not in terms of isolated concepts. They indicate that 'domains of experience' are structured wholes within recurrent human experience that are organised as experiential *gestalts*. These experiences seem to be *natural kinds of experience* in the sense that they are products of human nature (Lakoff & Johnson, 1980:117). Some of these experiences may be universal and others vary from culture to culture. In their later work, *Philosophy in the Flesh* (1999) 'domains' are not explicitly defined. In their view of 'cross-domain mapping' and 'radial categorisation' (or 'radial structure') (Lakoff & Johnson, 1999:177, 224, 225, 233, 500, 501, 512), one finds an indication of what this idea of 'domains' actually entails.

Lakoff and Johnson (1999:77) state that the empirical results of the work of second-generation cognitive science led to the abandonment of the key assumptions of Anglo-American 'cognitivism'. Two kinds of evidence were provided by this new research:

(1) a strong dependence of concepts and reason on the body and

(2) the centrality to conceptualization and reason of imaginative processes, especially metaphor, imagery, metonymy, prototypes, frames, mental spaces, and radial categories.

Basic-level conceptualisation is the cornerstone of 'embodied realism', they say (Lakoff & Johnson, 1999:74–93). Metaphorical language is a reflection of metaphorical thought. Metaphorical thought in the form of cross-domain mappings is primary; metaphorical language is secondary (Lakoff & Johnson, 1999:123). To this they add:

> [...] metaphors ground abstract concepts through cross-domain mappings using aspects of our embodied experience [...] Metaphors are the very means by which we can understand abstract domains and extend our knowledge into new areas (ibid. p. 543).

From this exposition it is clear that conceptual metaphor is deeply ingrained not only in the human body, but also in the human ability to interpret reality.

In the following chapters I will show that what makes this ability and characteristic trait of reality so unique is the fact that it not only harbours the vast multiplicity of potential meanings that could be refractively opened up through language, imagination and thought, but also that this refractive potential is open ended and ultimately pointing to the unity of meaning of all that exists and the wide spectrum of meaning that is locked into human existence and reality. This key role played by metaphor in cognition, thought and language is mirrored in the correlated analogical stratification of the world.

Part Two

Metaphorical Hermeneutics: A New Methodology

III Metaphors as Keys to the World and the World as Key to Metaphor

Metaphors and Experience

The argument that metaphor provides a unique avenue to open up both the seeing, interpretation and understanding of different aspects of reality and diverse angles of texts has become an acknowledged dimension of what is called *The Rhetorical Turn* in interpretation (cf. Gross, & Keith, 1997). Representatives of the hermeneutical tradition such as Gadamer (1989), Ricœur (1977b), Van Hoozer (1998), et al have developed important insights concerning the metaphoricity of language and the role of metaphorical models and language in the process of social scientific explanation. The term 'metaphorical hermeneutics' is associated with Gadamer's approach to hermeneutics. Here it is used in the sense of metaphors as keys to the world. Metaphorical Hermeneutics can be subsumed under the wider rubric of Rhetorical Hermeneutics. It includes both the weaker version of hermeneutics which concentrates on the way rhetorical figures are produced and used in discourse as such, and the stronger sense of hermeneutics where the discourse actually refers to thought and action and also gives access to ontological states of affairs (cf. Ginev, 1999:247–81).

Metaphorical hermeneutics is a theoretical approach to the interpretation of reality built on the notion that metaphors mediate in our understanding and interpretation of reality. The mediation takes place from two different angles. On the one hand it is a cognitive device which mediates epistemic access to reality. On the other hand it reveals something of the stratification of reality which correlates with the cognitive role of metaphor. Metaphor represents a new way of seeing that creates or opens up a new semantic field which is the result of the interaction between at least two domains or semantic fields. These 'domains' and semantic fields cannot purely be imputed to social and cultural conventions. The meanings ascribed to the metaphor-in-context are constrained by experience, which in turn is limited by the boundary conditions set by the structures or nomic conditions for reality. Various authors point to the fact that these universal structures do exist; that they condition human and societal behaviour and that it would in principle be possible to construct a theory of these structural universals or a 'biogrammar of the universal cultural acquisition device of humankind (cf. Harré, 1976:32; Tiger and Fox, 1974:17, 30). Cross cultural research in both psychology and anthropology have pointed to the existence of such traits, and in recent philosophical discussions Apel and others have pointed to the necessity of recognizing the existence of some sort of 'transcendentalia'. In the context of the discussions concerning the 'grounding' of the metaphorical utterances this question becomes extremely relevant. Snell (1953:220) for example states:

> We realize that the reality which (the 'necessary' metaphor) expresses lies far below the level of human or animal activity, that it is one with the very roots of our existence. This is the life that can be grasped intellectually, by the principle of the excluded middle (cf also Edie, 1975:41).

Apparently Ricœur refers to something similar when he requires a fundamental distinction between metaphor and symbol and states that metaphor '[...] testifies to the primordial rootedness of Discourse in Life' (1976a:59). Lakoff and Johnson make similar claims when they develop their theory concerning the relationship between metaphors and experience. These authors point to undeniable states of affairs recognized in different historical phases. The question posed by these insights is to what extent and how the metaphor (albeit aesthetic, religious or scientific in nature) touches or expresses this 'deeper' ontological reality and 'proper' analogy. I will argue that contemporary embodiment theories do not fully accommodate the relationship of embodiment to the correlate stratification of non-human life. I suggest that recognizing the existence of structural domains of meaning which condition the corporeal structure of human life and reality provides a more adequate basis for a theory of metaphor and analogy.

Such a theory will give an account of the basic domains of experience and their interrelationship, and will also have to explain the basis for the distinction between inherent and interactive properties which function in language use, definition of concepts and metaphorical language use. Literal meaning is as much context bound as metaphorical meaning, although the context is generally so much part of our background beliefs that we fail to recognize it as context (Kittay, 1995). This a differentiation at least requires some guidelines for the identification of contexts and the determination of 'improper' sort crossing or category mistakes. Hart's (1984) exposition of Dooyeweerd's ontology and his appropriation of Vollenhoven's notion of universality is very helpful in the clarification of these issues[1]. In this chapter I shall pursue some of the contours of a systematic philosophy which develops a view of human embodiment, human experience and domains of reality that provide important clues to the boundaries and limits or 'itineraries of meaning' (Ricœur, 1980) of metaphors and the unique function

1 The significance of such an ontology for the solution of problems concerning the relationship between metaphorical models and scientific realism has been dealt with elsewhere. (cf. *Proceedings of the Tenth Biennial meeting of the Philosophy of Science Association, USA,* 1986, vol. 1). Elements of this paper will surface in Part IV of this work.

of metaphors as 'keys' to the opening up of the meaning of the world.

Personhood: Relational and Stratified Embodiment

Metaphors always rest on some form of analogy which relates to differences and similarities of some kind or other. Metaphors are not only lingual articulations of purported interactive meanings but also typical of thought processes about the world. This is the case because they rely on ontic categories – provisionally called 'domains' – and in turn create new blends of meaning or 'mental spaces' (Fauconnier, 1994). Conceptual metaphors[2] provide conceptual frameworks which provide the scaffolding for the networks of meaning which language relies on and which different kinds of metaphor reveal. In the creative and interpretive interaction with the world, embodied experiential *gestalts*, image schemata and conceptual metaphors structure human experience. The way these cognitive abilities function point to the irreducibly stratified nature of both the embodied cognitive agent and the stratified world with which she is cognitively interacting. The challenge is to demonstrate how this irreducibly stratified reality constrains metaphorical meaning and meaning change and how it relates to the stratification of the embodied cognitive agent.

Embodiment indicates the whole, integral human person in his relationship to himself, others, culture, the mineral, vegetative and animal worlds and ultimately to what he believes to be God or a

2 At this point I use the terms 'conceptual metaphor', 'image schemata' and 'experiential gestalts' coined by Lakoff and Johnson. Later it will be apparent that there are differences in our understanding of these notions.

deity. The alternative approach suggested here counteracts subjectivism and materialism in the following ways:

- By recognizing the irreducibility of a vast multiplicity of entity and modal domains in reality and the irreducibility of different act structures within human embodiment it, avoids any form of reductionism.
- By acknowledging the radial structure of these irreducible modal domains, the intrinsic coherence of all spheres of reality are recognized. This in turn rules out absolutist claims that ground meaning only in one dimension of reality.
- By grounding metaphorical meaning in the universal, dynamic meaning stability of the irreducibile modal domains.
- By acknowledging that they are in turn conditioned by a design plan or order. It is to the contours of this purported design plan that we now turn.

The alternative view of the domains and radial analogies involved in experience and embodiment proposed in the previous chapter, is based upon ontological and anthropological distinctions found in the Philosophy of the Cosmonomic Idea of Herman Dooyeweerd (cf. *A New Critique of Theoretical Thought, 3 volumes,* 1953–57). This philosophical system develops an account of the coherence and unity of meaning of the diversity of structures found in reality. According to this theory of embodiment, human functioning always takes place within the context of the following relationships:

- a central religious relationship to God or what is regarded as a deity
- the relationship to one self
- the relationship to others

- the relationships to the non-human world
- the relationship to culture and history.

In these relationships human beings function as sexual and gender beings as individuals and in social relationships. This account sets the stage for an alternative understanding of the interactive nature of human knowledge and the inter-domain relationships to which metaphors refer.

Corporeal Structures

In the human body there are at least four interlaced corporeal structures, each having its own identity but also building on one another and depending on one another. They are

- the *physico-chemical* (inorganic) structure, based on the substrate of spatial, numerical, kinematic and temporal aspects;
- the *biotic* (organic) vegetative structure, based on the substrate of the physico-chemical aspects ;
- the *psychic* (sensitive) structure, based on the two prior substrate structures and
- the human *act structure* which incorporates the prior three embodiment structures and all normative dimensions of human existence. It is open ended in the sense that the acts it produces can be qualified in a multiplicity of possible ways.

These acts take place on the basis of this integral yet diversified embodiment infrastructure and are usually differentiated acts that potentially exhibit all the irreducible dimensions of reality and are normatively qualified in some way or other. All aspects of reality are always present in every act, but in each different context and activity they assume the colour of these guiding norms. In the act of driving a vehicle or caring for a family or writing a book, or painting a house, all four of these structures with their diverse characteristic aspects are present, but each act takes place under a different contextual set of guiding norms which qualifies the specific act. Reading and understanding traffic symbols is a 'language' of its own, as is the language of legal documents, cook books and sermons. Language functions in all these activities but in each case it is qualified and guided by a different set of contextual norms.

So speech acts and utterances are cognitive acts that are lingual, interpretative, interactive acts of the full human person in any one or more of the above mentioned relationships, under the guidance of contextual norms. Discerning the role of these contextual norms in the interaction between domains expressed by metaphorical language, requires an approximation of the underlying ontology assumed by the metaphor. So, although all the modal aspects of reality are potentially present in all entities, complex human actions always take place under the contextual guidance of qualifying norms which are characteristic of the context within which these acts take place[3]. These norms determine the semiotic configuration of the aspects involved. For example, knowledge of the stock market is guided by economic norms; knowledge of works of art is guided by norms for beauty, harmony, allusiveness; knowledge of butchering by techno-formative or economic norms of frugality, etc. The important question is what the grounding of the nature of these conceptual domains is. What are the pre-conceptual *experiential gestalts* (Lakoff and Johnson, 1980:117–118; 1982:193ff) that lie

3 Entities designate human acts, events, things, facts and societal relationships.

at the basis of conceptual metaphor? This has already been partially addressed in the previous chapter.

Human personhood comes to expression in an integral embodied existence which manifests more than what has traditionally been called the 'body' i.e., only the physico-chemical, biotic and sensitive dimensions of human existence. Moreover the diversity of dimensions characteristic of the human person points to it being centered in a diversity-transcending spiritual unity of existence. Acts performed by the human person always involve the full 'soul-bodied person' in whose activities one is able to discern a great diversity of aspects or facets and complex levels of structures. All human acts are enacted by the fully integrated person, not only by physical bodies, souls, minds or spirits. These actions take place within the limits of the constraints set by a great diversity of aspects, facets or functions characteristic of human life and reality. This notion of personhood overcomes the soul-body dichotomy, the subject-object divide, the body-consciousness opposition. It also grounds embodiment in the transcendental conditions that make existence possible. It postulates at least four different overlapping 'bodily' structures that presuppose one another and that are simultaneously involved in all human acts. Johnson (1991) refers to a similar state of affairs but calls it 'patterns'. Such acts take place within particular contexts which qualify them. Economic activity is not artistic activity (although there is always an economic dimension to good art and an aesthetic dimension to economic activity). Legal activities are not political activities (although there is always a political dimension to law and a legal aspect to politics). The meaning of metaphors utilized within these contexts obviously acquire their qualified meanings under the auspices of the context in which they function.

Each one of these four interrelated structures forms the substrate for the structure 'above' it. This 'hierarchy' implies that the physiological and neural grounding of metaphorical meaning is an essential and constitutive basis of all other forms of meaning. So e.g. linguistic meaning presupposes sensitive, organic and physico-chemical substrates. Social communication in turn presupposes the bases just mentioned and also human lingual and cultural abilities. Even though these constitutive bases can be discerned in all metaphorical construals of meaning, metaphorical meaning cannot be reduced to its material neural basis. Basic to the four above mentioned anthropological structures are the facts that

- All entities in reality and human experience function in a diversity of mutually coherent but irreducible aspects or facets of reality – also called modal or functional domains.
- This diversity of coherent but mutually irreducible modal domains exhibit both similarities-in-differences and differences in their similarities – also called modal or functional analogies.
- Analysis of the world around us rests on the identification and distinction of these similarities and differences exhibited by entities and/or aspects or facets of entities.
- Concept formation rests on the identification of such similarities and differences in which features that have been identified and distinguished are united in a concept.

The implications of these assumptions are that the quest for a literal grounding of meaning associated with the double-language thesis, is replaced by a grounding of meaning in non-analogical (or original) (Hart, 1984:161) modal meaning. This *non-analogical* meaning must be distinguished from analogical, modal meaning. These distinctions form part of a more encompassing ontological framework which claims to provide an account of the way structural domains of meaning condition metaphorical and analogical meaning.

Structural Domains of Meaning

The Amsterdam School of Philosophy of which the Dutch Christian philosopher of Law, Herman Dooyeweerd, is one of the founders, developed a comprehensive ontology which attempts to give an account of the rich diversity, unity and coherence of the irreducible aspects of human life, experience and reality. This philosophy starts from the assumption that every thinker, school of thought, tradition, theory and discipline harbours a view of reality which functions as transcendental presuppositions of everyday thought and also of scientific or theoretical thought. This view of reality gives an account of the diversity, coherence and unity of entities in reality. When we perceive, discern, conceive, analyze, articulate, argue etc., about the world we formulate concepts that depend on the distinction of similarities and differences that are characteristic of the irreducibly diversified and categorized world in which we find ourselves. Concept formation rests on identification and distinction of similarities-in-difference or difference-in-similarities in which we unite features distinguished and identified in a concept. Metaphors facilitate this process by enabling the selection of or disregard for similarities-in-differences or differences-in-similarities. The selection takes place on the basis of analogies. When metaphors are utilized in scientific theorizing they often facilitate the approximation of the structure of ontic analogies present in the diversity of reality. This means that metaphors are mediators to the provisional approximations of the structure of phenomena. This is why Hart (1984:159) says that metaphor can express a genuine connection – a functional analogy – not merely a semantic peculiarity. Although Dooyeweerd's view of modalities and functional analogies resembles Lakoff and Johnson's idea of 'conceptual metaphors' and radial analogies, these functional analogies are more than the conceptual metaphors postulated by Lakoff and Johnson. In order to highlight the difference between these functional analogies and conceptual metaphors we turn first to the role of metaphor and analogy in concept formation in Dooyeweerd's philosophy.

Dooyeweerd (1953) differentiates between different types of mutually cohering irreducible origin-*al* domains in reality and human experience. In each one of the irreducible aspects or facets of reality and of human experience the whole spectrum of other domains is mirrored and expressed. He writes:

> This universal character of referring and expressing, which is proper to our entire created cosmos, stamps created reality as meaning, in accordance with its dependent non-self sufficient nature. Meaning is the being of all that has been created and the nature of even of our selfhood. It has a religious root and a divine origin (1953:4).

Expressing points to the fact that the wide spectrum of the unity of meaning of reality comes to expression in the multiplicity of modal domains of reality. Referring alludes to the fact that the diversity of modal domains in turn harbour the potential of referring to a multiplicity of possible worlds. These facets or aspects of reality exhibit cross-domain analogical relationships which refer to the basic coherence of the stratification of the world. The consequence of this is the fact that all the basic concepts of the disciplines are per definition analogical. Strauss (2001:iv) distinguishes between metaphors and analogies on the following basis: Metaphors, he says, are lingual in nature and are entitiary analogies (e.g. atoms are like billiard balls) that pertain to similarities and differences between entities in reality, whereas similarities and differences between aspects or facets (modalities) of reality are indicative of modal analogies, also called *retrocipations* and *antecipations* in Dooyeweerd's philosophy. Retrocipations – referring to earlier modalities and antecipations – referring to later modalities in the hierarchy of modalities.

Each irreducible ontic domain is characterized by an original or non-analogical modal nucleus, kernel or irreducible core (Hart, 1984:157). This is the most basic common denominator of a range of family resemblances characteristic of that aspect of reality. One is tempted to say this meaning nucleus is the literal core, but the fact that this core expresses itself via analogical relationships rules out this kind of language use. A closer description would be that it constitutes the prototypical core meaning. For example, the origin-*al* meaning nucleus of the economic mode of reality could be formulated as frugality. Economy of thought or economy of speech are analogies between the economic and logical and the economic and lingual modes of reality. Even though they designate different states of affairs, they both appeal to the common original prototypical meaning of the economic mode of reality, i.e. frugality. The prototypical meaning nucleus of the spatial aspect is continuous extension and that of the aesthetic aspect is allusiveness, etc. (Seerveld, 1980:131). These meaning nuclei are the prototypical core meanings that qualify a range of family resemblances.

Every thing, fact, event, act and societal relationship functions in all these modalities and can be approached from any one of these modal angles. The active or passive presence of this aspect can be discerned in all of reality and human experience. The modalities or functional aspects '[...] form a modal framework in which we grasp concrete phenomena from different points of view' (Dooyeweerd, 1954). What makes this modal core or nucleus remarkable is the fact that it can only come to expression via a vast array of analogical elements. Thus in effect a multiplicity of potential and embryonic hierarchically ordered meanings are already present in the core or origin-*al* or non-analogical meaning of an irreducible aspect of reality (not only of human embodied experience). We come to know these meanings through human embodied experience. In these experiences metaphorical meaning transfer is mediated through the juxtaposition of two semantic fields. With these distinctions in mind we need to revisit the conventional distinction between metaphor and analogy.

The Conventional Distinction Between Metaphor and Analogy Revisited

Metaphors are used when an unfamiliar thing, event, fact, act or social relationship is being spoken of, depicted, portrayed, understood, interpreted in terms of another more familiar thing, fact, act or social relationship. Metaphor transfers meaning from a source domain to a target domain and in this process a new meaning is uncovered or 'created'. Metaphor is not only lingual in nature (cf. Botha, 1988a & b), but is characteristic of thought processes and ultimately lies at the base of most logical, psychological, imaginative human acts. Up to this point we have not registered any difference with the conventional view. Here our paths gradually diverge.

Metaphor indicates any statement or utterance which understands some concrete domain of human experience and reality in terms of another concrete domain of experience and reality. So statements such as 'education is gardening,' 'therapy is archeological excavation', 'parental love is a never ending stream', or 'the state is a system' are examples of metaphors relating two (concrete) entities, even though some of the source or target domains are actually 'abstract' in the conventional sense of the word. This distinction is based on one in which two kinds of 'domains' can be discerned: entitiary domains and modal or functional domains. An entity is not only a concrete physical entity but any concrete thing, fact, event, action or societal structure and relationship. 'Concrete' in this context refers to whole, integrated entities that can be acts, things, events or societal relationships. An election or an earthquake or a psychotherapeutic counseling session are examples of such concrete entities. A modality on the other hand is an aspect or facet of any such concrete entity. A metaphor is usually the juxtaposition of two concrete entities whereas an 'analogy' in the sense in which it is used in this philosophy, is a relationship between two facets or modalities of concrete entities. With this distinction in place the notion of metaphor becomes more refined and requires further specification when used within the context of the systematic distinctions of Herman Dooyeweerd's philosophy.

Entitiary metaphors often mediate modal analogies (cf. the reference to the different ways in which the metaphor of a 'system' can be used). Analogies that express the similarity-in-difference or differences-in-similarities between two entities are metaphors and are often mediated by and articulated in language. Not only is there an infinite number of potential analogies present in any such relationship, but some are already known as existing and recognizable analogical elements, whereas others are 'created' in the course of the opening up of the potential analogies present in the semiotic radius of the domains in which the analogy occurs. Analogies indicate a relationship between aspects, facets or dimensions (also called 'properties' or 'irreducible kinds of functionality' Hart, 1984:149ff) of concrete entities. As indicated earlier Clouser (1991) calls this *the principle of aspectual universality*. Examples of such modal analogies are 'economic growth', 'social distance', 'psychological stress', 'political movement', 'personal space', etc. These analogies are anchored to or moored in irreducible kinds of functionality which constitute the semantic domains that analogies reveal. Lakoff and Johnson's notion of the existence of a large system of primary metaphors comes close to this idea. Conceptual metaphors are often conventional, they say. Lakoff and Johnson differentiate a set of metaphors they call 'ontological metaphors' which are the type of metaphors where abstract notions are thought of as concrete entities or substances[4].

It is important to differentiate clearly between the standard usage of the term metaphor as found in poetry, prose and literature, and the more sophisticated way in which it is often used in metaphor theory where it indicates deeper ontological, orientational and structural states of affairs. The latter type of metaphor inevitably incorporates some form of ontic analogy. In the following section I show that the basic analogies which metaphors often uncover are embedded in and indicative of the structural irreducibility and coherence of reality.

'Home Bases' and Radial Analogies

These modal analogies come to expression in the elementary basic concepts of disciplines. They rest upon the irreducibility and mutual coherence of modal aspects. In the sciences and academia we cannot rid ourselves of these metaphorical and analogical usages (Hart, 1984:153). Every analogical concept we use is 'at home', has an address (Brandt, 2000 in a different context) or is original in a certain domain of reality and is used in all the other domains. This requires further explanation. Let us assume that the 'home base'

4 The term 'ontological' is also used in the more conventional philosophical sense of the word when Lakoff and Johnson analyze Aristotelian categories and essences. This use of 'ontological' differs from its use in the Philosophy of the Cosmonomic Idea.

(original/non-analogical) aspect comes to expression in an unlimited number of possible analogical moments. When a concept is used in its ('literal') non-analogical sense, it qualifies and determines the nature of all the analogical meaning concepts used within that theory or discipline. So e.g. when physics uses the term 'force' in the context of a mechanistic understanding of reality, this 'colours' the meaning of all other terms utilized within the context of that specific theory.

The nature of an analogical concept is that in it is present all the references to earlier modalities, differently formulated; these aspects are already constitutive of the modal aspect. These meanings are 'per definition' present within the frame of reference or the meaning of the specific term or concept. This simply means that when one is referring to physico-chemical entities one's concepts already incorporate spatiality, time, numerosity and movement. The concept does not yet necessarily include biotic, logical, lingual, social, formative, juridical, ethical or fiduciary elements. Those could become incorporated by analogical extension. All disciplines deal with space, time, motion, organic life, psychic feeling, lingual codes, signs and symbols, cultural development, symbolic meaning, economic evaluation, moral phenomena, juridical issues, etc. But in each discipline the typically analogical concepts are prefaced with some qualification[5].

So for example, mathematics (geometry) deals with space as analogical concept. The spatial mode has a meaning nucleus which guarantees its *original* meaning ('Pure spatiality' i.e. space in pure 'unadulterated' form). Physical space can not be reduced to mathematical space. Historical examples of approximations of such an idea of space in its unadulterated form are Newton's idea of 'absolute space' or Kant's idea of the intuition of space (in the

5 This analysis is based on Dooyeweerd (1954).

original sense). Some of the analogical meanings associated with this original meaning are:

- The *space of sensory experience*, seeing, touching, hearing which is different from pure mathematical space.
- *Personal space* is a psychological phenomenon qualified by sensitive, emotional laws and not by space.
- *Gravitational fields* are not spatial entities dependent on spatial or physical continuity. They are not perceptible to the senses, but are 'fields' of a different kind. They are fields of energy – kinematic phenomenon/energetic phenomenon.
- *Logical thought space* is analytical in character. Logical extension of concepts and judgements are not spatial extensions but truly logical in nature.
- The concept of a *cultural area* which is characterized by the historical/cultural/formative notion is not a physical space but historical or cultural areas are defined not by space, but by historical or cultural power, e.g. Western civilization or Third World Civilizations are typically culturally qualified concepts.
- A *language area* is a super-sensory phenomenon which is unbreakably related to sensory perception but derives its meaning from the fact that the 'area' (space) is qualified by being of a lingual nature.
- The market is an *economic area*, which is related to physical and geographical space, but qualified by economic laws (Carrying coals to Newcastle).
- 'Legal area of competence' (*Jurisdiction*) is qualified by juridical norms (Flag of a ship as symbol of the legal jurisdiction of a specific country.)
- *Social space*. Social is home base – space/distance is an analogy referring to spatial)
- *Kinematic* space.

- *Liturgical* space (an altar)
- *Ethical* space (intimacy)
- Multiple other examples can also be noted, such as *economic space* / stock exchange/ world markets.[6]

A few other examples should suffice. When feeling or sensitivity and movement are seen as two irreducible modes of functioning, then the following are examples of an analogy. A 'moving experience' is one in which the psychic dimension of feeling and sensitivity relates analogically to the dimension of movement. In music on the other hand an 'adagio' refers to slow musical movement. In the latter case, there is an analogical relationship between the aesthetic dimension and the dimension of movement. In both cases movement functions but in each case a different dimension of movement or analogical moment of movement is being utilized. The concept movement, which is presumably an original conceptual 'metaphor' exhibits multi-ordinal (Korzybski, 1933) meaning in its relationship to other irreducible dimensions of human functioning. The following two diagrams show that the facets of reality are multi-ordinal and exhibit cross-domain analogical relationships which refer to the basic coherence of the stratification of the world. This means that all the basic concepts of the disciplines are per definition analogical. What follows from this is the fact that wherever the original notion of space is dependent upon a non-spatial qualification, this concept bears an analogical relationship to the original concept of space (Dooyeweerd, translated by Knudsen, 1954:3). This theory of metaphor and analogy will be dealt with again in later chapters and will be implemented in the analysis of the meaning change that takes place in the development of the *force* concept in Michael Faraday's work.

6 See Figs. 1 & 2 of radial analogies of the psychic (psychological) and spatial domains.

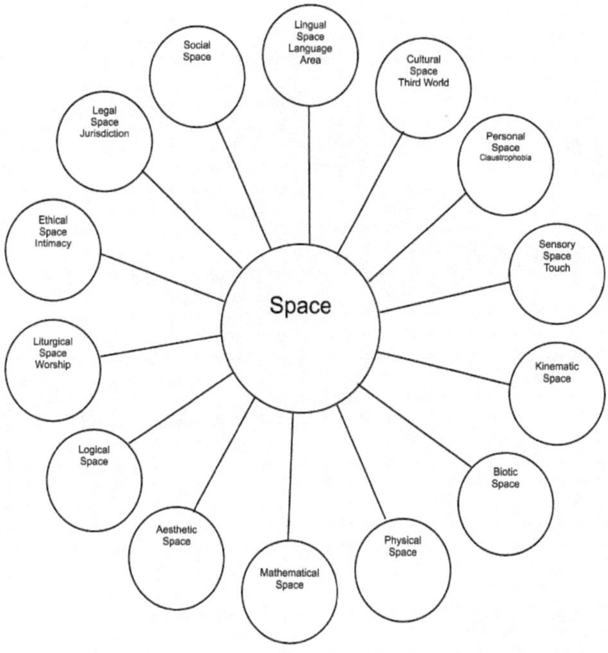

Figure 1. Radial functions: Space domain with analogies

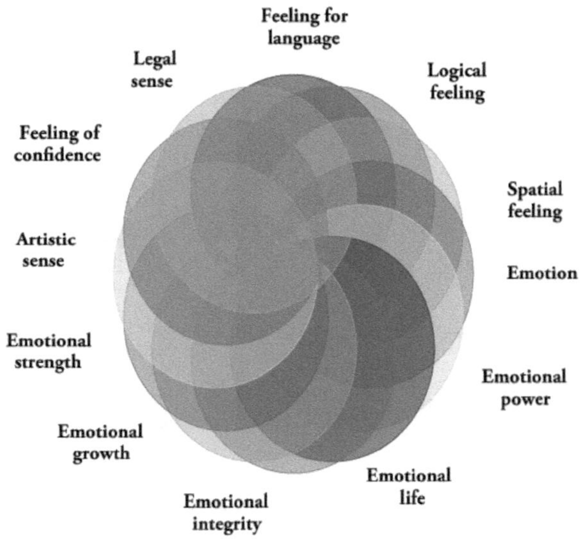

Figure 2. Radial functions: Psychic (Emotional) domain with analogies

All Metaphors Are Analogies But Some Are More Analogical Than Others

Although metaphors usually express some sort of analogy, metaphor should clearly be differentiated from a certain type of analogy which is not conceptual in the first instance but ontic. 'Ontic' means pertaining to the general structure of the world. An ontology would be a theoretical articulation of such a structure. My provisional agreement with Lakoff and Johnson and others about the 'conceptual' nature of metaphor rests on an important caveat viz. that this conceptual structure which lies at the basis of linguistic articulations of metaphor, is grounded in a deeper analogical ontic structure of the world and of human experience. It is the 'metaphorical' (actually 'analogical') ontological structure of this grounding and the role of the focal analogies in the process of metaphorical meaning transfer that is of interest here.

It is apparent that all metaphors are analogies, but some seem to be more analogical than others. The irreducible diverse, yet coherent ontic structures which condition meaning and reference are not metaphorical in nature, but they are analogical. The concepts indicating these irreducible diverse analogical structures are unique and original and cannot be replaced by literal paraphrase or substitution. Analogies are relationships between modal aspects of reality. Examples of such modal structures are spatiality, extension, number, physical force, motion, vitality, distinction, signs/codes, retribution, harmony, frugality, troth, faith etc. These aspects exhibit a nucleus or core meaning. When this core meaning is unpacked it inevitably happens through the articulation of the analogical moments of a modality. We can only come to know what justice is by tracing the analogical relationship of the juridical modality to other modalities. Principles of good faith, equity, good morals, legal power, legal judgements are examples of such analogical moments (Dooyeweerd, 1953:29).

All disciplines utilize basic analogical concepts in their theories about the nature of reality. These are concepts that the discipline absolutely can not do without. Examples are 'society' in the case of the social sciences, emotion or sensitivity in the case of psychology and faith in the case of religion and theology. In each discipline these concepts function in a typically qualified way. Physical force, psychological force, economic forces, social force, cultural power etc., all make use of the term *force*, but in each discipline and in the theories in the discipline the concept is imbued with a different hue. The variety of qualifications which these concepts undergo in various disciplines is ultimately bound to the original or non-analogical modal meanings of these notions. If these were metaphors they could easily be replaced by other concepts or words or combinations of terms without their specific original and non-analogical significance. Such 'metaphorical constructs' are original, primitive or non-analogical. One should not even call them 'metaphors' because of the fundamental difference between these two notions. The analogies are pregnant expressions of the coherence and interrelatedness of all of reality.

Analogical relationships between entities are often expressed in metaphors, for example 'love is a journey' (Lakoff and Johnson, 1980:25). Such metaphors refer to concrete entities, are primarily lingual in nature and can be substituted by other metaphors. They refer to similarities and/or differences between entities, whereas ontological and functional analogical relationships between aspects of entities are of a different kind. They are unavoidable analogical structures (Hart, 1984:156) that reflect the topography of domains of reality. Metaphors often mediate our access to these unavoidable analogical structures. They mediate access to these ontic analogies through the process of analogical concept formation. Analogical concepts express the colour of the field within which they are used and are also often the conceptual manifestation of the basic metaphor, root-metaphor or world view operating in a discipline or theoretical perspective. I shall demonstrate this in the analysis of the different ways in which Faraday uses the concept of 'force' in his physics. We shall now proceed to differentiate the conventional understanding of metaphor and analogy from a deeper type of ontological (ontic) 'metaphor'.

Necessary Metaphors and Proper Analogies

Basic to all metaphors is the fact that they rely on some form of analogy. This is especially the case in science where it is practically impossible to function without these analogical concepts. The analogical concepts are more than mere figures of speech and semantic peculiarities (Dooyeweerd, 1954). Within the tradition of Reformational philosophy there has been a strong resistance to using the term 'metaphor' to designate these analogies and analogical concepts, yet Hart's (1984) valuable 'translation' of the Dooyeweerdian notion into the contemporary terminology concerning metaphorical language has been extremely helpful.

The important distinction between metaphorical analogy and ontological or 'proper' analogy, is a distinction inherited from the classical theory of metaphor. In the classic theory of metaphor there was always a reticence to equate metaphor with what was regarded as 'proper analogy' (Burrel, 1973:260). Metaphors point to specific types of analogical relationships that are of such a nature that they cannot be ignored. They have to be acknowledged. Moreover the literature provides ample reference to and evidence of the fact that some metaphors are actually more 'necessary' than others (Ortony, 1975:45–53). 'Necessary metaphors' (Snell,1953) differ from metaphors found in everyday language and also from those used in poetry, scientific explanation and in religious contexts. They also differ from other semantic peculiarities such as idioms and figures of speech. As such they are presupposed by all the other above mentioned types of metaphors. Edie (1975:39) too, refers to the distinction between *epiphors* and *diaphors,* but then argues that diaphors are more fundamental, root, and 'necessary' metaphors which are frequently not recognized metaphors at all because of their absolute fundamental function of organizing experience. Edie's view seems to confirm the possibility that a certain type of metaphor refers to an ontological level. In order to clarify some of the distinctions discussed so far, the relationship between metaphorical language and the 'necessary' metaphors or 'proper' analogies which some types of metaphorical language are able to discover, needs further discussion.

Necessary metaphors are pointers to the unfolding of the meaning potential of modal spheres and prove to be vehicles which enable scientists to probe the fundamental analogical structure of reality. In both everyday knowledge and science it is not possible to rid ourselves of the analogical or metaphorical usages which might be called functional metaphors (Hart, 1984:156) or ontological 'metaphors'. These metaphors which refer to the fundamental or ontological (it would be more accurate to refer to 'ontic' or ontical' in this respect) analogies found in reality distinguish themselves from other metaphorical usage in the sense that they designate actual states of affairs. When one refers to the 'growth' of the economy, it points to something present in the economy that can only be expressed analogically. The economy does in actual fact behave as though it is growing and expanding, whereas in the case of a mere figure of speech the apparent incongruence responsible for the tension in the metaphor does not refer to actual behaviour. Expressions like 'man is a wolf' or 'Richard is a fox' do not imply actual behaviour of this kind. When one attempts to avoid these 'metaphors' in scientific theorizing, it soon proves to be impossible. Hart (1984:153) writes: 'It is not possible to express realities in any functional mode without use of terms that seem borrowed from or that seem to refer to other functional modes'. Hart (1984:160) poses the question whether the semantic metaphor is not a pointer to ontic analogy, i.e. that the likeness between economic and plant growth, length of time and length of lines, the scope of a piece of land and the scope of vision, etc. point to ontic analogies which represent more-than-semantic reality. Hart is here pointing to a phenomenon that other authors have called 'necessary' metaphors. The basis for this phenomenon is the *principle of aspectual universality* (Clouser, 1991:215).

The Principle of Aspectual Universality

Every human act (thing, fact, event, entity, societal relationship) in principle exhibits all the modal aspects or facets either passively or actively. In each one of these dimensions all of the rest of reality is reflected or echoed. A possible image to portray this process of referring and expressing of the aspects of reality is the image of a prism breaking up unified white light into the spectrum of rainbow colours. One could also speak of the 'focal hues' of white light. But the principle of aspectual universality implies a significant modification of the prism image, namely that each of the hues in turn also reflects, echoes or mirrors the others. To some extent one could speak of an infinite series of refractions. The important principle here is that reality is coherent and unified and yet shows its diversity of colours in the process of meaning diversification and meaning disclosure. This meaning diversification is open-ended and infinite.

In analogical concepts the preceding aspects (substrate modalities) are assumed and subsumed. One cannot say *space* without implying that this concept also entails number/countability. Anything that occupies space must of necessity be countable. But space does not necessarily entail movement or sensitivity – those are properties that are 'higher/later' in the modal scale and require opening up. Movement and sensitivity are passive and potential modal aspects that can be opened up. One cannot use any concepts that refer to movement without implicitly assuming or implying both space and number. Things that move occupy space and can be counted. These are necessary conditions (cf. Hart, 1984:159). One cannot use the concept *social* without assuming the implicit presence of and reference to analogical elements in other aspects. It presupposes space, discrete quantity, countability, movement, logically discrete identity and difference, communication via language formulated in concepts, sentences, historical/cultural formation, etc. So when a metaphor mediates an analogical concept it often expresses already existing resemblances between two ontic and semantic fields. The type of analogy that mediates the creative and imaginative opening up of as yet not acknowledged or recognized aspects of reality and human experience actually refers to aspects that are 'higher' (later) in the modal scale (Stafleu, 1980:26). The distinction between *epiphors* and *diaphors* or even Johnson's (1980:55) distinction between the *canonical and non-canonical* metaphors show a close resemblance to this type of distinction. Implicit in such concepts are also the possibilities of interactively opening up passive aspects or potential qualities that are dormant in other subjects or objects. These projections are related to human experience and possibilities, i.e., perceivability, quantifiability, economic exploitation, ethical evaluation, etc. But they are properties of the objects or subjects and not only qualities that the knowing subject ascribes to them.

In this respect Dooyeweerd '[...]adheres to a realist interpretation of what empiricists called secondary qualities' (Glas, 2002:148–72). Active functions are those that are constitutive of the entity's existence and functioning. Human beings can not exist without actively functioning in all the dimensions above. Passive or potential functions are those that are dormant and can be opened up e.g. the economic functions of gold ore, or the aesthetic function of gold (jewelry) or the ethical function of gold (wedding bands) as sign of ethical troth. The last three examples demonstrate the way in which dormant/passive aspects of a rock are interactively opened up by human (normative) intervention, but the properties are passive and dormant in the piece of rock itself. The peculiar characteristic of the aspectual nucleus is the fact that it represents a colour of an aspect of reality and in turn reflects the full scope of the inter-modal coherence of the irreducible aspects of reality. Lakoff and Johnson would attribute these potential and dormant functions which conceptual metaphors can actively open up, to the human embodied experience. Obviously that is true, but the fact remains that these non-human entities intrinsically possess these potential qualities and or properties that can be un-covered through human experience. The multivocality of causality is a good case in point.

Causality in its origin is a physical phenomenon, but can be seen from a great diversity of vantage points. From a legal perspective legal causality (who is legally to blame) implies an analogical relationship (or cross-domain mapping) between the physical and juridical domains. Ethical causality (who is morally guilty), and social causality (social responsibility e.g. an impaired driver), represent cross-domain mappings (analogies) between the physical and the ethical and the physical and the social respectively. This example can be expanded to include economic causality (costs of the accident), etc. The radial meaning diversity within the physical domain thus shows an original or non-analogical (physical) and analogical use of concepts which is demonstrated by the multivocality of causality. Obviously such analogical refractions of a core origin(-*al*) meaning reflect a multiplicity of possible domains and cross-domain mappings, and these domains are in turn grounded in the experience of embodied persons. One important dimension of this analysis needs to be brought into perspective. This is the role that the focal analogical moment plays in the configuration of analogical moments in a metaphor.

The Role of Focal Analogical Moments in Metaphors

The focal analogical moment which is often the 'point' of a metaphor is embedded in the semiotic radius and/or semantic fields of two interacting ontic domains. So for example one can discern different semiotic and semantic fields related to the two entities system and organism. A system might for example be qualified by the fact that it is seen to be either mechanical or organic. A specific modality (which is characterized by a prototypical nucleus) will colour and qualify each one of the interacting fields within the metaphor; e.g. organisms are biotic phenomena whereas a system may be mechanical or cybernetic, etc. When the semantic fields of the two entities interacting in the metaphor are 'blended' in a metaphorical construct the 'point' of the metaphor is to lift out or abstract a specific focal element of the source domain in order to highlight a specific element of the target domain. A metaphor in which for example an organic system, like a plant, is utilized in order to understand the target domain, like a political society, would accentuate the biotic analogical element in the relationship, whereas a mechanical system metaphor accentuates the mechanical analogy. To use mental space theory here, one could say that the 'blended space' which comes to expression in the organic system metaphor differs from the blended space resulting from the use of the mechanical-system metaphor. In both cases 'system' was invoked but in each case a different aspect of the system was accentuated and used as a 'filter' or 'screen'. The main reason for this is the fact that in the actual input space ('system') of the two metaphors the (analogical) moments mentioned are configured differently because of a different analogical focus.

A critical question that is raised by these phenomena is what exactly in reality and human experience makes this diversity of meaning possible? Answers to this question vary, but the literature shows a strong proclivity towards constructivist approaches. That human construction of metaphorical meaning is instrumental in this process needs little argument, but the question remains as to what it is that limits and constrains these itineraries of meaning (Ricœur, 1980) to which metaphors point. This issue becomes even more important in scientific theorizing.

IV Metaphors As Keys In Science

Changing Views of Language, Metaphor and Science.

The central thesis of this chapter is that the resources of metaphor are essential to the work of science and that the construction and retention of metaphor must be seen as part of the aim of science. It is therefore argued that metaphorical models are constitutive elements of scientific theorizing. They attempt to theoretically approximate the nomic structures for (and of) reality by means of hypothetico-structural explanation. Metaphorical models provisionally approximate some aspect of this complexity and provide access to functional analogies characteristic of reality. Metaphorical language often plays a pivotal role in these models and constitute linguistic and semantic 'tools' which facilitate new ways of seeing reality and provide unique and novel possibilities for the redescription of the explanandum. Some metaphors are more fundamental and basic than others. They point to what has traditionally been called 'proper analogy' (Burrell, 1991).

The tension or suspension of disbelief generated by metaphor in ordinary language is also characteristic of models based on metaphors. The statements 'light is a particle' or 'light is a wave' are incongruous to some extent and reveal this tension. When postulating a metaphor that atoms are billiard balls, an incongruence is created which comes to expression in a certain tension within the metaphor. Although the tension can be attributed to linguistic or semantic ambiguity or incongruence, it can also be seen as an apparent reflection of the integral coherence of various aspects of reality which resist the artificial *epoche* of theoretical analysis. In spite of this tension in the metaphor it is able to mediate a blending of meaning and of semantic fields which is partially the result of the interaction between the juxtaposed semantic fields. Although the tension present in a metaphor and in a model based on a metaphor can ostensibly be attributed to linguistic or semantic relationships, the argument in this chapter is that the tension generated by using a metaphorical model to access the structure of reality is the result of the complex coherence of reality which resists the theoretical *epoche* brought about by the application of the model. Both literal and metaphorical language rests on similarities-in-difference that reflect the categorization of reality. Metaphor is a unique tool which creatively and stereoscopically combines elements from these categories and in doing so mediates ordinary and theoretical concept formation. The tension present in metaphorical language is an essential element of the process of knowledge acquisition in general and also of scientific knowledge acquisition. In both the acquisition of knowledge in everyday experience and in theoretically mediated scientific experience, the characteristic juxtaposition intrinsic to metaphorical language signifies a relationship between different categories fundamental to all human experience. Theoretical metaphorical language mediates access to structural states of affairs that are characteristic of the complex coherence of reality. The use of this type of language is a significant part of what McMullin (1978) calls hypothetico-structural explanation. This understanding of the nature of theorizing is a more recent development in Philosophy of Science. It was preceded by a view of models and metaphors that was deeply embedded in Aristotelian views of language and positivist views of the nature of theories (Johnson, 1991:3–47).

The groundbreaking work of Black, Hesse, Kuhn and others emphasizing the role of metaphors, analogies and models in scientific paradigms, paved the way for understanding that metaphorical language is not ornamental and therefore not redundant in theorising. On the contrary it has come to be seen as constitutive of many scientific models and moreover able to claim cognitive import. In science as the hypothetico-structural explanation of the structures for reality, metaphorical models provide both the focus of the theory as well as the language with which a theory attempts to explain some dimension of the complexity and coherence of reality. As Rothbart (1984:610–11) indicates, metaphor is instrumental in conceptual problem solving, which is at the heart of theorizing. In this chapter some of the recent developments in Philosophy of Science that have appropriated metaphor theory for a better understanding of the nature of scientific theorizing, are introduced. The central thesis here is that the resources of metaphor are essential to the work of science and that the construction and retention of metaphor must be seen as part of the aim of science.

The traditional reductive theories of metaphor – the simile and substitution theories – have been replaced by the interactive theory of metaphor which has led to an elaboration of the traditional classical distinction between *epiphors and diaphors* and to a better understanding of the innovation which transpires through the use of metaphor. The recognition of the role metaphor plays also has implications for the question concerning the purported discontinuity between pre-theoretical and theoretical language and concept formation This interactive understanding of metaphor is a later development which differs fundamentally from the traditional Aristotelean and positivist understanding of metaphor.

The Interaction View of Metaphors and Scientific Models

The positivist treatment of metaphor that prevailed during the first half of the twentieth century is actually a version of the traditional centuries-old empiricist critique of metaphor. According to Johnson (1981:17) the positivist's critique rests on two foundations: the first is the distinction between the alleged 'cognitive' and 'emotive' functions of language, and the second is the attendant belief that scientific language could be reduced to a system of literal and verifiable sentences. These two tenets were conjoined with the earlier empiricist literal-truth paradigm which argued that metaphor deviates from literal or cognitive discourse and if it had any cognitive import at all it would be expressible by literal statements of comparison. The positivist tradition also fostered the well known so called reductive theories concerning the nature of metaphor, i.e. the simile and the comparison theories, which both attempt to reduce metaphorical expression to either a substitution of the one referent of the metaphor by the other or to a comparison of the two referents. In both cases the effect is a reduction of the metaphorical to the literal similarities or analogies.

Black's (1962) interaction view of the nature of metaphor challenged most of the basic tenets of the reductionist views which had become entrenched in the empiricist myth. In Philosophy of Science the challenge to the empiricist myth came from various sources. Both Hesse's (1966) and Kuhn's (1973:202) emphasis on the pivotal role of language in models, and scientific descriptions were certainly significant contributions to an alternative understanding of the structure of science and the nature of scientific language. These challenges collectively led to a radical rethinking of the nature of scientific language and a fundamental critique of and an alternative to the reigning tenets of the empiricist account of the nature of science. The seminal ideas instrumental in much of the discussions, concerning metaphorical language have been that of Richard's *The Philosophy of Rhetoric* (1936) and the attempted explication and elaboration of Richard's view concerning the 'interanimation of words' by Black in his well known *Models and Metaphors* (1962) and his *More About Metaphor* (1979:19ff). The seminal ideas of Black have been elaborated by a whole list of authors, including Johnson, Turner, Lakoff, Rohrer and. Fauconnier. The decisive contribution of Black's theory of metaphor to the existing body of knowledge concerning the function of metaphorical language, is the fact that it provided valuable new insights into the dynamic nature of language and the creative process of meaning disclosure and meaning variance.

These insights made it possible to criticize and relativize the standard view of metaphor which had become part and parcel of a static view of language with its emphasis on the so called 'double language thesis' (Beardsley, 1967:287), i.e. the distinction between literal and metaphorical language use. Black not only criticized this view, but proposed his alternative interaction view, which opened up new avenues to understand how the disclosure of novel meaning in language and science takes place through the formation and use of metaphor.

Black argues that the metaphorical utterance works by projecting upon the primary subject or system of the metaphor a 'set of associated commonplaces' comprised in the implicative complex that is predicable of the secondary subject or 'system' of the metaphor (1979:28). These associated commonplaces are not private to individual language users but are largely common to a given language community and are presupposed by speakers who intend to be understood in that community. They are part of the language for that community. Black claims that the interaction brought about by the metaphor has the effect of changing both the primary and the secondary systems. In this dynamic semantic interaction some new structural correspondence between two metaphorical subjects is expressed. This means that language is dynamic and constantly changing by reciprocal shifts of meaning. The appropriateness of the metaphor depends at least in part on some principle of assimilation between primary and secondary systems, variously described as 'analogy', 'intimations of similarity' or 'a framework through which the primary is seen'. The implication of this is that the interaction view cannot consistently be made to rest on an initial set of absolutely literal descriptions, but rather on a relative distinction of literal and metaphoric in particular contexts.

Black also extended his explanation of the function of metaphor to cognition and science. This approach was also developed and elaborated by Hesse and McMullin. It is especially in Black's later publication (1979) that he recognizes a closer bond between models and metaphor and between metaphor and analogy and claims that '[...] every metaphor may be said to mediate an analogy or structural correspondence' (1979:31). This means that the metaphor functions as a principle of selection, emphasizing certain traits or facets of the primary subject and suppressing others. Metaphors may refer to very concrete entities or aspects of entities, yet when they are utilized in scientific models and theories[1], they mediate the process of abstraction through which certain structural similarities and differences come in to view. This view of theorizing is in strong contrast to the classical view of science and scientific language.

When Black's interaction view of metaphor is applied to scientific models, the 'two systems' present in the metaphor should be interpreted as referring to aspects of reality signified by two appar-

1 Cf. Mellor's (1968) discussion of Duhem and Campbell's views concerning analogies, where a similar point is made.

ently unrelated semantic fields. These semantic fields are linguistic representations of two dimensions of reality. In the case of scientific metaphors the so called primary and secondary systems may be highly organized networks of theoretical laws (Hesse, 1966:160). The metaphor focuses the attention on some element of structural correspondence between two fields which becomes visible or comprehensible only in the cross section of these two perspectives. In the tradition that Black represents, metaphor is seen as central to the task of accounting for our perspectives on the world: how we think about things, make sense of reality and set the problems we later want to solve. Metaphorical utterances are significant expressions of a particular kind of 'seeing as', the 'metapherein' or 'carrying over' of frames or perspectives from one domain of experience to another (Schön, 1979). This brings about an interaction which causes a meaning shift to take place, one in which the primary system is 'seen' differently. It brings about a form of vision which Stanford (1936:105) very aptly calls 'stereoscopic vision'. The exact nature of this interaction, the conceptual displacement and conceptual innovation which is brought about and the manner in which metaphorical description mediates the disclosure of meaning and actually approximates the unknown features of reality, form the core of a cluster of problems dealt with extensively in discussions of Black's position. The growth of this understanding concerning the 'interactive' process of meaning transference has led to greater clarity concerning the role of metaphoric models in hypothetico-structural explanation.

Both Kuhn's account of the structure of science and Hesse's modification of the empiricist position in defending the role of metaphors in theorizing, presupposes the *underdetermination thesis.* Quine succinctly formulates the thesis as follows: theories that are under determined by data are '[...]theory formulations that are empirically equivalent, logically incompatible, and irreconcilable by reconstrual of predicates' (1975:322). The underdetermination thesis is the claim that for any theory for a given subject matter, there is a presupposed incompatible rival theory which is evidentially equivalent. It was realised from an early stage in the scientific revolution that there is a logical problem about the inference of theories from observations, because these cannot be made logically conclusive: there are in principle always an indefinite number of theories that fit the observed facts more or less well. A scientific theory is always under determined by the facts it purports to explain. One set of facts can support a variety of theories. The implication of the acceptance of the underdetermination thesis is the fact that in principle a multiplicity of possible redescriptions of the explanandum are called for and this now justifies the fact that the same state of affairs can be described by means of a diversity of metaphorical descriptions.

The primary significance of Hesse adopting Black's interaction view of metaphor in her network model of theories (1966, 1970, 1972 and 1974) and in her network model of meanings (1983) is the fact that it leads to a fundamental modification of the hypothetico-deductive account of scientific theorizing and a relativization of the traditional logical positivist distinction between observation language and theory language. The understanding of the role metaphorical models play in scientific cognition has been stimulated by recent developments in metaphor theory in general and more specifically by insights gained through the gradual displacement of the traditional double language thesis by the understanding of the inherent metaphorical character of all language[2]. These insights have made it possible to relativize and modify the observation-theory distinction inherited from the hypothetico-deductive account of scientific theorizing, and replace it with a recognition of

2 Perhaps speaking of the categorical nature of all language is closer to what is meant here.

the theory ladenness of all observation. In this theory ladenness metaphorical models play a decisive role, both in a heuristic and in a constitutive sense. The hidden structures, which condition the existence of the observable phenomena studied in scientific disciplines, are approximated by theories. The theories in turn are developed on the basis of models which facilitate conceptualization and conceptual problem-solving. This process constitutes some of the most fundamental problems of all theorizing. Ricœur's (1977b) contribution to the understanding of the dynamics of metaphorical language has contributed immensely to a better understanding of the significant role of metaphorical utterances and devices both in everyday language use and in science. The constitutive role of metaphorical language in scientific models provides an important point of access to the mediating role of metaphor as vehicle for metaphysical and religious convictions.

Scientific Theorizing as Hypothetico-Structural Explanation

Scientific theorizing and explanation is directed towards the mapping of the structures characteristic of the explanandum. Scientific models developed on the basis of analogies generate metaphorical descriptions of reality which provide an extension of the scientific vocabulary. The existing scientific vocabulary is the sedimentation of previous scientific theories and as such the observation language of such a scientific description is theory-laden in terms of the categories and classifications built into the existing theoretical language system of the field. The theoretical language which is utilized in the description of the facts is thus not necessarily the same as the proposed theory put forward to explain these facts (McMullin, 1968:388). An explanatory theory is intended not as a description of what one already has, but as a hypothesis, something that goes beyond the evidence by introducing a postulated structure that could provide a causal account of the data to be explained, although it is not itself directly observable. The postulated structure is the model, whereas the theory is the set of statements in terms of which this structure is provisionally explained. The theory is derived from the model (McMullin, 1968:389). McMullin (1978:139) calls this type of explanation hypothetico-structural and states that theoretical models provide us with a metaphorical and tentative idea between the unknown and the observed properties. Such a postulated structure or model functions as a hypothesis concerning the explanandum. In this respect it is important to distinguish between models and theories and the role that metaphors play in them. Soskice's (1985:102) distinction between 'theory-constitutive' metaphors, i.e. metaphors which propose a model, and metaphors which are linguistic projections of such a model, is helpful in this respect. In order to differentiate the two types of metaphors, she suggests a useful distinction between *theory-constitutive metaphors* and 'metaphorically constituted theory terms'.

Harré and Martin's helpful description of the relationship between a model and metaphor illustrates the point: When the image of a fluid is used '[...] to explicate the supposed action of the electrical energy', '[...]the fluid is functioning as a model for the conception of the nature of electricity' (1982:100). When, however, '[...] we speak of the 'rate of flow' of an 'electric current', they argue, we are using metaphorical language based on the fluid model. The model thus gives rise to a 'spin off' of a matrix of terminology which can then be used by the theorist as probative tool.

Speaking metaphorically on the basis of a model, a scientist is enabled not only to posit but to refer to theoretical entities by the use of terms which transcend experience in that their semantic context is not fully determined a priori by the empirical conditions for their application (Harré and Martin,1982:101,102). The metaphor provisionally provides a mediation to fill the existing terminological hiatus. I would like to add that the matrix of terminology mediates the analogical moment (movement of fluid) which acts as the focus of the model.

Theory-constitutive metaphors acquire their heuristic function from the fact that they actually provide access to what is often called the 'hidden structures' behind phenomena which they claim to explain. Certain types of metaphors approximate deeper level ontic structures – the fundamental or proper analogies referred to in the previous chapter. One could say these are the hidden structures of reality or the 'joints of the world' which scientific theories attempt to explain by developing explanatory models based on analogies and formulated in metaphorical language. The matrix of terminology which a model provides, functions as a fertile source for hypotheses and suggestions which aid the process of scientific theorizing or hypothetico-structural explanation. According to McMullin (1978:139) the function of hypothetico-structural explanation is to facilitate understanding of observed properties or behaviour of the explanandum and to discover its intrinsic structures. A theory could be seen as a set of propositions explanatory in its intent. Where the explanation is a structural one, part of the theory will be a specification of the structure. The structure underlying such an explanation is often called a theoretical model, since the explanation is a hypothetical one. The entities constituting the structure are often called theoretical entities.

Structural explanation needs to be distinguished from nomothetic and genetic explanation. In nomothetic explanation one appeals to an empirically determinable regularity or 'law', whereas genetic explanation attempts to answer questions concerning the origin of a particular kind of structure or behaviour (McMullin, 1978:143). Structural explanation is more powerful than nomothetic explanation. Structural explanation involves *retroduction*, whereas nomothetic explanation involves *induction* as a type of inference. Empirical laws reached through the method of inductive generalizations have rarely been supposed to conclude with certainty to a true and definitive formulation (McMullin, 1978:145), and do not necessarily introduce conceptual innovation or conceptual displacement. Retroduction on the other hand, introduces conceptual novelty. The leap involved in retroduction is far greater than in induction in which there is a leap of extension from the particular to the general[3]. The justification for the introduction of the retroductive hypothesis is its explanatory success. Explanatory power is thus the grounds for the sort of ontological claim that retroductive inference allows one to make (McMullin, 1978:145, 6). The historical success of hypothetico-structural explanation is best explained by supposing that the existence of the theoretical entities and lawlike relationships between them, referred to in the explanation, is highly confirmed by the success of the explanation. McMullin concludes that this is one of the strongest arguments in favour of scientific realism. The notion of hypothetico-structural explanation provides a viable account of the nature and structure of scientific theorizing. Before the implications of this view are discussed, the role of analogies in theoretical concept formation and structure mapping warrants a closer look. Here both the term structure and analogy will be used in the way it is used in Reformational philosophy.[4]

3 The term 'abduction' introduced by C.S. Peirce has a similar meaning.

4 The term 'structure' can be used in two different senses: It could refer to the factual or subject-side of reality. Subject here means being subjected to the normative or lawlike requirements or structure of reality, e.g. gravitational phenomena are subjected to the laws of gravity. In this sense 'struc-

Dooyeweerd does not pay much attention to metaphors or the role they might play in either the discovery of analogies or mediating theoretical models and theories. On the contrary. His comments concerning metaphor and the function of metaphor in theorizing are predominantly of a derogatory nature. References to metaphor in Dooyeweerd's philosophy are sparse and usually made in the context of the discussion of the fundamental role played by analogies (Dooyeweerd, 1955:55,64). One such a reference is in Dooyeweerd's discussion of Stoker's analysis of the so called 'substance concept' (Dooyeweerd, 1969:68ff). Metaphors, Dooyeweerd then argues, can be the expression of the aesthetical imagination. They have no other role to play than to evoke a visionary picture of nature (ibid.1969:68).

In his discussion of the analogical use of the fundamental concepts in the different branches of science, he points to the fact that the relation of analogy – as he understands it – must be investigated within the cadre of the modal structures of meaning, which are determined by this order (1969:58–9). In the description of this state of affairs the lack of uniform terminology in the different languages and the linguistic ambiguity of words that may also have a metaphorical sense, present a real problem. In his discussion of the original and analogical use of the term 'space', he enumerates

---

ture' designates the structure *of* reality. It could also refer to the law-side of reality, i.e. the structural laws governing and conditioning the regularities perceived in phenomena. Then it is understood in the sense of the structure *for* reality. When the term 'structure mapping' is used in this chapter it refers primarily to the latter sense of the term. Although it is conceded that analysis of the structures *for* reality are inextricably bound to the analysis of the structures *of* reality.

the different ways in which this term is implemented in various disciplines and then states emphatically that the variety of qualifications which this concept undergoes in various disciplines is ultimately related to the fundamental meaning-moment of space, extension. Then he proceeds: 'If it were a metaphor, the term in its scientific use could simply be replaced by another word or by a combination of terms without any spatial significance' (Dooyeweerd, 1969:64–65).

In his article *De analogische grondbegrippen en hun betrekking tot de structuur van den menselijken ervaringshorizon* (1954:171) he denies that these fundamental analogical concepts are in any sense metaphorical in character. Metaphors are seen as ambiguous use of terms, whereas analogies refer to the ontological coherence of the diversity of meaning. The significant role that metaphorical constructs do play in the process of uncovering these analogical moments is not dealt with or utilized in Dooyeweerd's account of the nature of science – his theory of the *Gegenstand*srelationship. He did not appreciate the role of metaphor in theorizing and contrasted the 'visionary world of the poet' and the imperative faced by the philosopher to explain the exact theoretical meaning of the concepts he uses. This may point to the fact that he accepted the 'double language thesis' with its distinction between literal and metaphorical language use. Most probably it should be attributed to various other factors, such as inter alia, the fact that he regarded analogies primarily as ontic states of affairs and clearly distinguished them from metaphors which he regarded as typically lingual in nature. Although he did not refrain from employing metaphorical language in his own systematic philosophical exposition (the image of the prism for the notion of time for example) he advocated a definite discontinuity between ordinary and theoretical language. Metaphor would be relegated to the realm of the language of pre-theoretical experience whereas the language of theorizing would be regarded as far more exacting both in precision and in meaning.

With respect to the multidimensional meaning of the analogical concepts of the special sciences his concern was the recognition of the underlying ontological structures which conditioned their nature and meaning. These irreducible structures, he argues, are not metaphorical in nature, because the concepts employed to designate them are unique and original and cannot be replaced by literal paraphrase or substitution (Dooyeweerd, 1955:64–5). From these references it seems as if Dooyeweerd held some sort of reductive theory of metaphor, one in which it would be possible to either replace the metaphorical expression by another expression or else reduce it to some literal paraphrase. The point he wants to make in this argument is not one concerning the nature of metaphor in the first instance, but one concerning the nature of the ontic analogies, which are foundational to all language in casu metaphorical language. The innovations introduced into metaphor theory by Richards and Black et al and which generated the 'interactive' theories of metaphor, most possibly were either not known to him or else not regarded as relevant to the problems he was addressing at the time.

Dooyeweerd's notion of the role of the religious groundmotives in science, culture and history resembles some versions of the notion of a root-metaphor (Pepper, 1970) as found in many writings on metaphor. In Dooyeweerd's analysis these groundmotives come to expression in basic scientific models and are at the root of insights that have been overemphasized or absolutized. In this respect the organistic and mechanistic models utilized in contemporary naturalistic science are good examples. It is often argued that these models are actually overextended metaphors, yet once the significant relationship between metaphors and the ontic states of affairs they attempt to articulate becomes evident, then it is clear that more is at stake than only a lingual category mistake or extension. What actually takes place is a fundamental distortion of some specific ontic state of affairs. Although this is an important point of contact with developments in recent metaphor theory, it can not be dealt with here.

Metaphors: Stereoscopic Contexts for Analogies?

In both everyday knowledge and science it is not possible to rid ourselves of the analogical or metaphorical usages. This means that a distinction has to be made between unavoidable analogical structures (Hart, 1984:156) or proper analogies and their linguistic articulation by means of metaphors. The fact that analogies lie at the heart of a theory is well established (Harré, 1978:171; McMullin, 1974; 1984:30). What is argued here is that it is not just any analogy that will fulfil this function, but that it is primarily the *fundamental analogies* which are uncovered by the fruitful metaphor of a fertile theory. The concrete analogies from which a theory often originates, provide the heuristic access to the hidden structures which science attempts to uncover in its retroductive attempts at structural explanation. These hidden structures are approximated by the discovery of the fundamental analogies. A basic idea concerning the nature of these fundamental analogies is at the heart of the conceptual apparatus which a theory utilizes to explain reality (Stafleu,1980:26).

Conceptual novelty which provides some epistemic access to hitherto unknown domains and facilitates successful structure mapping, is an articulation of a whole series of expectations about the manner in which certain aspects of the world do behave or function. The systems, objects or items covered by a certain new concept share overlapping similarities. In the case of scientific concept formation these 'family resemblances' are instances of 'functional analogies' (Hart, 1984). They point to basic and underlying ontic analogies. In the approximation of these functional analogies scientific imagination is the vehicle for creative opening up of novel insights into the structure of the world and also for the disclosure of new meaning. It is impossible to ignore these 'metaphors' in scientific theorizing.

The well known examples of analogies within Reformational Philosophy illustrate the fact that within one multifunctional functor there exists an interrelatedness which can come to expression in two different ways: It either expresses 'inherent' qualities or else 'interactive' properties which point to the interrelationship between various functional dimensions. Apart from the intermodal analogical relationships representing the semantic interaction between two aspects present in one functor[5], it is also possible that similar phenomena present themselves in the so called subject-object relationships, i.e. the relationships between functors. As argued earlier this implies that entities can exhibit resemblances and differences, but also that properties of these entities can resemble one another and yet also differ. Even when a metaphor does not express the relationship between two concrete entities (the state is a mechanism) but the intrafunctional interrelatedness within one entity (psychological growth), the functional analogies often implicitly presuppose the concrete phenomena in which modalities are present. In a phenomenon such as *psychological growth* there is a latent underlying resemblance between human personalities and other 'growing' phenomena, such as plants, trees etc. Psychological growth might refer to the functional interrelationship between two

5 See list of analogies discussed under the principle of aspectual universality.

functions within one person, but it still presupposes two distinct modal aspects known to us in their concrete manifestations, often in disparate phenomena. This is the result of the fact that it is often difficult to visualize realms of functionality or modalities without referring to concrete phenomena of which they are the functions. It is clear that this phenomenon refers to something more than a mere comparison between human personalities and growing entities. It provides a biotic filter or screen through which psychological phenomena are seen and as such provides a novel perspective, in which stereoscopic vision opens up the possibility of seeing new facets that cannot become visible from a single perspective.

There exists a close interrelatedness between the above mentioned 'intrafunctional analogies' and the analogies which become apparent in so called 'subject-object relationships'. The most elementary form of the latter relationship is the fact that the majority of natural entities of the macro world are all visible, audible, touchable, etc. (Hart, 1984:240). This would most probably be the type of qualities that Lakoff and Johnson (1980:121) have in mind when they claim that concepts are at least partly defined by interactional properties having to do with perception, motor activity, function, purpose, etc. This, they claim, is the basis for the fact that our concepts of objects, like our concepts of events and activities are characterizable as multidimensional *gestalts* whose dimensions emerge naturally from our experience in the world. When such an analogy is utilized in a scientific model it provides a syntax to construct a transposable, but original discourse. This is not a mere subsumption of one mode of description to another. In the case of subsumption the fundamental concepts of the subsumed semantic field are unchanged. When metaphor is utilized it leads to a metaphorical variation of the key concepts of the subsumed theory. Metaphors thus often function as the vehicles of the fundamental analogical concepts of the specific sciences. This becomes apparent when a revolutionary scientific innovation introduces a new root-metaphor into the scientific terminology of a discipline.

The argument developed above points to the fact that the metaphors utilized in the development of scientific models provide access to underlying structures which condition the explanandum. The terms and concepts within the limits of a semantic field related to and determined by the limits of a specific mode of functionality can be used in two different ways: They can be utilized within the conceptual limits of the modality within which they function originally, i.e. in a constitutive sense (Strauss, 1973:178). They can also be used in an adherent or interactive sense, in which case the concepts refer to dimensions or functions which exceed the limits of the semantic field to which they originally belong. Metaphors of the interactive type fulfil a suggestive and creative function and often give rise to conceptual and semantic innovation. Dooyeweerd's (1955:186) distinction between concept and idea is an important point of departure here. Epiphors could be equated with the use of metaphorical concepts in the retrocipatory or 'concept' mode. This would mean that epiphors embody modal retrocipations which function in the process of modal abstraction. The reason why the interactive type of metaphor is fertile in theory innovation is because it retroductively provides hypothetico-structural access to possible avenues and structures of reality which can be elaborated through research, theory construction and theory validation. It is a source of possible hypotheses that can be explored. Metaphorical models based on analogies are therefore instrumental in the approximation of the hidden (analogical) structure that scientific theories discover.

Metaphor and Theoretical Concept Formation

Concept formation is basic to scientific theorizing and to the displacement of concepts which takes place in scientific revolutions. Metaphor is often the vehicle through which these conceptual transformations take place. It fulfills this function in different ways: It can be a mere heuristic device or be utilized in an instrumentalist fashion or it can function in a constitutive way in the process of theorising. What distinguishes a theory-constitutive metaphor from heuristic and instrumental metaphors is the fact that it proposes to replace one obsolete or less successful metaphorical theory as metaphorical redescription of the domain of the explanandum. The proposed alternative metaphorical redescription claims to possess more problem solving potential or is in actual fact more fertile in the suggestion of new avenues of research (Hesse, 1980:111ff).

The unavoidable ambiguous and multivocal nature of many basic concepts used in theories has led to much interest in the metaphorical and analogical nature of these concepts. Analogical concepts which are found at the heart of theories exhibit some puzzling traits: Some 'metaphorical' constructs apparently are unavoidable in theorizing and indicative of analogical states of affairs in reality which intrinsically harbour the potential of multidimensional meaning and reference. These puzzling states of affairs are attributed to the analogical nature of the underlying pre-conceptual and pre-linguistic ontic categories which determine the boundaries between analogical and non-analogical meaning.

Some constructs '[...] sweep through the whole array of scientific disciplines' (Maasen, 2000:217) and cross over the borderline into everyday discourse. *Chaos* is an example of such a concept. When one unpacks the meaning of the concept in the variety of conceptual meanings that it is found, it points to the absence of order or lack of equilibrium. Both these phenomena can be seen as originally physical phenomena. When one tracks down the diversity of meanings that this term acquires in a variety of theoretical contexts, it becomes clear that theoretical frameworks determine the meaning career of such a concept. Chaos has been seen as the opposite of order, the source of order, the bearer of hidden order etc. (Maasen, 2000:214–15). How does the term acquire this wide diversity of meanings? Disciplinary differentiation provides a partial explanation. Each discipline zeroes in on a different angle of the phenomenon. But, within the disciplines the different theoretical approaches sharpen the focus even further. Ultimately it is the focal analogy and control beliefs within a theory that determines which aspects or facets of reality come into focus. In this process the guiding analogical moments of the theory-constitutive metaphors play a significant role.

Even though concrete metaphorical images or concepts are used as central theoretical constructs in theories (e.g. the systems concept), it is one or more of the aspectual or modal facets of such a concrete metaphor that often functions as the controlling, guiding, steering element that directs, constrains and limits the meaning and reference of these terms. The guiding character of such an analogical element needs to be distinguished from the qualifying function of guiding norms of acts as distinguished earlier . Liturgy is qualified by norms of worship, a family by bonds of troth. But when the 'family' metaphor is used to describe a state or a school it is one characteristic element of the family metaphor that functions as the focal and controlling element – the decisive element in the grid of the screen or the size of the filter. Theory-constitutive metaphors use concepts that are grounded in irreducible aspects of reality. The analogies characteristic of these irreducible aspects of reality provide the basis for the mapping of the structures for reality.

Analogies and Structure Mapping

A condition for the possibility of the process of structure mapping with the aid of metaphors is the existence and presence of analogies. A metaphor based on an analogy can map the structure between a known domain and a domain of inquiry. Structure mapping of analogies assert that identical operations and relationships hold among non-identical objects. Such an analogical relatedness is a prerequisite for metaphorical descriptions. Often a metaphor approximates such a purported analogy by expressing or articulating the contours of the existing analogical resemblances which characterize the analogical relationship. The structure mapping that takes place is not only a process of mapping *between theories* pertaining to theoretical domains of inquiry, (e.g. Darwin's appeal to the uniformitarian principles of Lyell's geological theory and to Malthus' principle of natural selection) but it can also take place between *world and theory* (billiard ball model for the behaviour of gases) or *theory and world* (Bernoulli's discovery of the correct equation for the flow of water from a reservoir by 'seeing' the water flow as an exemplification of pendulum motion, [Bradie, 1984]).

The analogy expressed in a metaphor or a metaphoric model refers to relational structures between the two systems. Such an analogy requires overlap in relations which are necessary for the perception of similarity between two domains. Gentner (1982:111) claims that 'overlap in both object attributes and inter-object relationships is seen as literal similarity, and overlap in relationships but not objects is seen as analogical relatedness'. So e.g. in physics the known characteristics of water waves are projected upon sound or light (Hesse, 1966:11). Another good example is Harvey's use of the novel metaphor 'circle' to explain the movement of the blood in the human body. Harvey's circle hypothesis replaced the conventional wisdom of the times originally developed by Galen that one kind of blood ran from the liver through the veins to all parts of the body to perform nutrifying functions, while a different kind of blood performed a vivifying function (Rothbart, 1984:611). This hypothesis of continual circulation required a significant reformulation of the concept of the movement of blood. It provided a solution to the problem that William Harvey faced viz.: How could the heart pump out more blood in the space of one hour than the weight of a person? The high rate of movement of blood through the human body could only be explained through the metaphor of circulation. This leads to a displacement of concepts of the older theory and an *aggrammatization* into the language of the discipline, contributing directly to the conceptual problem-solving process of developing a coherent theoretical explanation of the laws governing the phenomenon i.e. the structure of the movement of blood (Rothbart, 1984; Schön, 1963). The analogy between the problem (the movement of blood) and a circle provided a clue to the solution of the problem. But not all properties and characteristics of the analogue function in this process of metaphorical transference. The philosophically interesting question concerns the identification of those properties or characteristics which actually do refer to the structure regulating the flow of blood and the reasons why specifically these traits are emphasized in the semantic field represented by the metaphor.

Initially the solution to this problem was sought primarily in the presence or absence of resemblances. This was the point of departure of reductionist theories of metaphor. Black rejected this approach and emphasized the so called set of associated commonplaces characteristic of both referents of the metaphor. Recent developments in linguistics in the so called semantic field theory provide significant insights into the dynamics of this process of meaning shift. It is interesting that Hesse uses the metaphor network to explain this process, whereas Kittay and Lehrer (1981) the metaphor field and in a similar fashion Rothbart (1988:388) the term 'vector'. The dynamics of conceptual displacement which takes place through the metaphorical interaction characteristic of both ordinary and scientific language, can be adumbrated through a closer look at the role of semantic networks in language and in science.

Semantic Networks in Language and Science

In the interaction view it is argued that literal meanings are shifted by their association with the metaphor. An expression initially metaphoric may become literal (a dead metaphor), and what is at one time literal may become metaphoric. What is important is not to try to draw a line between the metaphoric and the literal, but rather to trace out the various mechanisms of meaning shift and their interactions. In this regard Hesse writes: 'the interaction view cannot be made to rest on an initial set of independent literal descriptions, but rather on a relative distinction of literal and metaphoric in particular contexts' (1972:249). Hesse argues that what holds for metaphorical language in ordinary language use, namely that it is affected by the contextual interaction between the various elements present in the metaphor, is also true of scientific models. They function in a similar fashion (1972:254). She does not reject the distinction between literal and metaphorical language, but claims that the distinction is a pragmatic one. The distinction concerns the way in which speakers learn, use, and if necessary, define the words of their language. The commonly accepted distinction between literal and metaphorical or analogical meanings Hesse claims, is as baseless in all descriptive language as is the distinction between theoretical and observation languages in science. '*All* descriptive terminology is learned and subsequently used by a process of extension of application of meaning from similars to similars' (1976:8). This process of extension and application of meaning takes place in a similar fashion in science. This means that scientific description includes a continuous spectrum from elementary examples such as colour predicates, through less trivial metaphorical extensions of ordinary language, to the coining of new, but immediately understandable theoretical terms such as electric charge, electric current, etc. to the most esoteric terminology of modern physics. The consequence of this approach for the understanding of theory in the natural sciences is a much more flexible model of the interactions of the observable and that which explains it, i.e. the theory (Hesse, 1976:8).

The fact that metaphorical redescriptions of the primary domain of the explanandum in science leads to a deepening of understanding and an extension of insight into certain structural aspects, must be attributed to the new way of seeing which the metaphor makes possible, she says. This not only constitutes a figurative or symbolic manner of speaking about the primary subject, but actually refers to and truly describes some aspects of the structure of the primary domain. The consequence of this in scientific language is that the meaning of the original literal language of the primary system is changed by the introduction of the metaphor. The dynamic changes in theoretical language represent the changes of the intended meanings of the scientific community and become apparent in the meaning shifts that take place in the interaction between the two referents of the metaphor. This leads to an extension of our understanding of the world and an extension of the scientific vocabulary. It is thus one of the most significant cognitive mechanisms by which scientific knowledge grows. The contextual qualification of language which modifies and relativizes the double-language thesis in ordinary language functions in a similar fashion in scientific language. In an analogous way the contextual qualification and modification of the distinction between observation and theory language also takes place in scientific language.

The application of the interaction view of metaphor to the understanding of scientific language formation leads to the conclusion that the traditional observation-theoretical term dichotomy is not as radical as it was claimed to be in the 'standard view of science'. The distinction between observable and theoretical terms need neither be absolute nor universal. The observable-theoretical distinction can be drawn anew within each science as each new theory is incorporated into the existing body of scientific knowledge. Mellor illustrates this point well when he states:

> To be an observable in a science is to be referred to by a term so enmeshed in the science's law network that its correct application to relevant experience (controlled observation) is taken to be unproblematic, direct, not involving interpretation or assuming any particular structure of relations between it and other such terms (1968:288).

Hesse would argue that this comes about as the result of the gradual modification of the network of meanings of the observational terminology of an existing established theory (read: scientific description and explanation) by the (new) theoretical concepts which function as analogical extensions of this observational terminology. The interaction process between observational and theoretical language in metaphorical redescription can be elucidated by the process of conceptual displacement in hypothetico-structural explanation (McMullin, 1978). McMullin (1984:35) claims it is the characteristic trait of suggestiveness which provides the avenues for possible modifications and extensions of the theory, whereas Rothbart (1984:610) provides important clues to the relationship between metaphor, concept formation and conceptual problem solving.

Conceptual Displacement

Rothbart (1984) says the cognitive gain from metaphor arises within the process of solving conceptual problems in science. These problems are related to difficulties encountered in the structural explanation of unknown or anomalous phenomena. Conceptual problems often pose some of the most significant problems in scientific progress. These problems arise externally if a given theory clashes with some other strongly held theory or principle, or internally if a given theory violates canons of consistency, clarity or cogency. A theory's effectiveness in overcoming internal or external conceptual problems is a measure of its epistemic status. Conceptual problem-solving is an essential dimension of scientific rationality. Metaphor leads to concept reformulation. Concept formation takes place around members of a domain sharing some prominent features with some prototype members. This leads to a contextual classification of objects. Objects are grouped according to some shared prominent features, which in turn are determined by the object set (Rothbart, 1984). Rothbart argues that metaphoric projection '[...] reorganize(s) the semantic field by introducing new saliencies into the field by highlighting some features and eliminating others' (1984:610–611). This leads to the formulation of new attributes '[...] that can be directly beneficial when a conventional field of concepts fails to permit certain desirable features to emerge' (ibid.). Assuming that a conceptual problem is some weakness within the system of concepts, the gain from metaphor is expansion of the range of possible features attributable to the subject. 'When the primary subject is juxtaposed with prototypes from an alternative field, the metaphoric projection causes a reformulation of the network of similarities and differences' (Rothbart, 1984:611). This view is shared by Hesse (1983). It is the nature of these prominent features and new saliencies revealed or created by the metaphor, that requires further attention. The intriguing aspect of this process is the question concerning the manner in which the displacement of concepts (Schön, 1963) takes place.

Conceptual novelty which provides some epistemic access to hitherto unknown domains and to successful structure mapping, is an articulation of a whole series of expectations about the manner in which certain aspects of the world do behave or function. The systems, objects or items covered by a certain new concept share overlapping similarities. In the case of scientific concept formation these family resemblances are most probably instances of functional analogies (Hart, 1984). In the approximation of these functional analogies scientific imagination is the vehicle for creative generation of novel insights into the structure of the meaning of the world and also for the disclosure of new meaning; meaning which often is not already contained within the semantic network of the language systems utilized to conceptually formulate the anticipated similarities. A closer look at a few examples, provided in the work of Schön (1963:157ff) illustrates the argument developed above.

In the history of personality theory the notions of 'system', 'motions' and 'forces and fields' are borrowed from the scientific vocabulary of physics to explain psychological functioning. These 'forces' are different from Newtonian forces. The terms are imbued with meaning transferred from the dominant Maxwellian theory of physics which replaced that of Newton (Schön, 1963:156). The terms were borrowed from the physics of the time and can be attributed to the dominant root-metaphor of natural science which functioned indirectly as a projective model for new theories. The above mentioned terms plus terms such as 'dynamic', 'field' and 'energy' pervade the writings of *Gestalt* psychologists like Kohler and Koffka, dynamic social theorists like Lewin and dynamic psychiatric theorists like Sullivan, Horney and Fromm-Reichman (cf. Sullivan, 1964; Lewin, 1959). These writers have displaced to the mind the dynamic field physics of the late nineteenth and early twentieth centuries. In them the physics of Maxwell has replaced the Newtonian mechanistic view of forces (Schön, 1963:156). Schön (1963:157ff) gives an excellent exposition and comparison of the different ways in which Bentham and Lewin are influenced by the models provided by respectively Newtonian physics and Maxwellian physics. This becomes quite clear in their divergent use of the term 'force'.

In the gradual development of the concept of 'force' as a theoretical construct it was imbued with changing connotations within different theoretical contexts (Nersessian, 1987). Analysis of the shifts in the conceptual structure of the force metaphor in Michael Faraday's work shows it to be representative of a transition from a mechanistic to a dynamistic understanding of matter. Faraday's injunction and strong emphasis on the literal reading of the two books of nature leads him to a dynamistic understanding of matter and of reality. Dynamism is the view that all phenomena of nature, including matter, are manifestations of force. Such a view of force could be described as 'metaphoric hypertrophy'. In all three examples mentioned the term force is used metaphorically. It illustrates that our everyday experience of the world is dependent on the notion of 'force' with its related notions of energy and dynamism, but also that this basic experiential notion can be imbued with diverse scientific connotations.

This multivocality of the concept 'force' comes to the fore in ordinary language expressions of a metaphorical nature, such as: 'the force of an argument' or 'the movement has spent its force[...]'. In the everyday language expressions the metaphor of 'force' can be replaced by any other without doing violence to the meaning of the expression. 'Force' is a multivocal or multiordinal characteristic of human experience both in the pre-theoretical and in the theoretical sense. This multivocity becomes apparent on the level of theoretical concept formation and the displacement of meaning that takes place when one theoretical paradigm is replaced by another. This seems to be a phenomenon closely related to what Lakoff and Johnson calls the 'radial structure' of a concept, e.g.,

> The various theories of causation proposed over centuries by philosophers reflect the multivalent radial structure of an ordinary concept such as causation which is not monolithic but has an inherently metaphorical sense (Lakoff and Johnson, 1999:224–6).

This is reminiscent of a trait that Wittgenstein called 'polysignification' and which he regarded as crucial to the logic of metaphor (Gill, 1996:4). It is clear that in all these cases one can not do without metaphor. Whereas in everyday language and perhaps even in poetic language one can replace one metaphor with another, in the case of metaphors used as explanatory devices this is not necessarily the case, mainly because if they are apt, fitting and explanatory they actually refer to deeper ontic realities which can only be expressed analogically.

These examples point to the fact that natural scientific concepts are utilized 'metaphorically' in social scientific theorizing and often provide significant avenues to understand the behaviour of social phenomena. It also illustrates the interrelatedness of the various dimensions of reality and the necessity to determine criteria concerning the contexts in which these concepts are legitimately employed. Their presence does not necessarily point to theoretical distortion, but they do present the potential for such distortion. Not only does this cross-contextual utilization of concepts point to the fact that the development of the various scientific disciplines are closely interrelated and to some degree the expression of the same scientific world view, but it is possible that the opening-up process in one discipline (Stafleu, 1980:21) might be instrumental in the implementation of a similar process in other disciplines. The 'opening-up process' is the progressive unfolding of aspects of reality through the use of theoretical models based upon new analogical viewpoints.

Scientific Metaphors: Qualified by Logic, Language or Imagination?

Metaphors do not only express analogical relationships between concrete entities but more specifically between aspects of entities or relationships between aspects of entities. They mediate in the process of cognition. In the case of scientific metaphors they mediate in the process of disclosure of *structures* for reality. The contemplation of likenesses includes both an imaginative perceptive moment (to *see* or *seeing-as...*), a moment of logical identification and distinction (logical concept formation) and the linguistic formulation of such a concept which leads to a restructuring of semantic fields. The ability to perceive such likenesses harbours at least two distinct facets, viz. identification and distinction or the recognition of sameness in spite of the differences (Ricœur, 1978:148). This is most probably what Hesse had in mind when she argued that the family resemblances are not *said* but *shown*. This apparently refers to a pre-linguistic ability such as imagination. Ricœur (1978:145) speaks of the 'semantic role' of imagination in this respect.

The role of image schematic structures and imagination is crucial to the process of meaning disclosure. Theories concerning the nature of imagination and its role in cognition vary. Most philosophical positions are heavily indebted to Kant when it comes to theories of imagination. Johnson argues that '[...] *there can be no meaningful experience without imagination*, either in its productive or reproductive functions' (1987:151). According to Johnson imagination is the chief means for establishing order in our experience. He writes: '[...]our ordered world, and the possibility of understanding any part of it, depends on the existence of this synthesizing activity' (1987:151).

Johnson's view is helpful but we need to at least distinguish between the pervasive human ability to intuitively understand the order for the world and for our experiences and the foundational human ability to imagine this order. Imagination in scientific theorizing is a complex process which includes this fundamental ability of *seeing-as[...]* It recognizes existing similarities, facilitates their formulation and suggests the possibility of other likenesses or similarities between systems. In this process the metaphor mediates the transformation of perception and focus by facilitating stereoscopic vision. Perhaps it is exactly this apparent ambiguity which provides the basis for both arguing that theories could either be *logically* qualified artefacts or *lingually* qualified artefacts. Language functions as vehicle in the process of theoretical analysis and yet metaphorical models in fact represent more than language; they function as intuitive approximators of similarities-in-difference. What characterizes the effective scientific metaphor is its ability to offer possibilities for suggestive hypotheses concerning the structure of the actual behaviour of the phenomena under consideration. This leads to retroductive concept and theory formulation and language plays a crucial role in this process. Yet, the nature of concept formation is in fact the synthesis of a multiplicity of moments in a unity and as such logical abstraction, identification and analysis is presupposed.

What is important here is the fact that the interactive process includes more than mere perception. It is perception-opened-up by creative imagination, logical thought and most probably also linguistic categories. It is a 'seeing-as[...]' experience in which '[...] the flashing of an aspect on us seems half visual experience, half thought' (Wittgenstein, 1958:197). 'This insight into likeness', Ricœur claims, 'is both a thinking and a seeing. It is a thinking to the extent that it effects a restructuration of semantic fields; it is trans categorical because it is categorical' (1978:147,48). The important tension built into this process of seeing-as, is the fact that it includes both seeing x as y and recognizing that x is not y. The metaphorical 'is' which articulates the 'seeing as', signifies both 'is like' and 'is not'. It is clear that the verb 'to be' which figures in a metaphorical description cannot be identified with the verb 'is' when it is used in a literal description of reality, Ricœur argues. This distinction between perception and the 'more-than-perception', operating in a genuine diaphoric metaphor, is most probably also the basis for the distinction between the comparative type of metaphor and the interactive type of metaphor, and also the basis for the possibility of distinguishing these two levels within one metaphor. Mere perception of similarities as such provide the basis of the former, whereas the latter requires an opened-up type of perception.

Conclusion

The methodology applied in the next chapter assumes that both natural and social life are characterized by structures that are intrinsically 'metaphorical' (actually 'analogical') in nature and that can be approximated by means of theory-constitutive metaphors and metaphorically constituted theory language and models. Important empirical work in this regard has been done by the cognitive historical approach to science as developed by Nersessian (1987) Tweney (1989b) and others.

When using metaphorical hermeneutics as methodology to interpret, explain, describe or understand some aspect of reality, a text or a social action, the focal analogical moment present in the stereoscopic semantic fields of the metaphor provide hermeneutical access to some dimension, aspect or facet of the structure of the explanandum. For example when society is either seen as an organism or a mechanism or a system, the different emphasis in the configuration of analogical moments in the specific metaphor mediates the interpretation or explanation of the explanandum. Such a model is a deliberate construction on the part of the theoretician who attempts to approximate the structure of the text or action. To what extent such a purported structure is merely a conventional construction or actually approximates the nature of the text and its underlying structure is a matter of dispute in postmodern realism discussions and discussions of the indeterminacy of the text in literature. The grounds for this hermeneutical function of metaphor are to be sought in the structure of reality, human embodiment and the nature of cognition. Our exploration assumes that metaphors depict and refer to actual states of affairs in reality and have cognitive import. They depend on domains of meaning, conceptual domains, domains of discourse and semantic fields.

Part Three

Metaphor, Religion & Cognition

V Embodiment and the Fiduciary Dimension of Reality

Religion in Science

Metaphorical hermeneutics functions as methodology in the analysis of the role metaphors play as keys to the understanding of the world. It also provides a methodology to access their role in the theoretical understanding of reality. In this chapter a third dimension of the hermeneutical role of metaphors is uncovered viz. the mediating role that metaphors play in the influence of religious convictions in scientific theorizing. Views on the nature of religion and on what constitutes 'religious' commitment vary, as do views differ on whether and how religious convictions influence scientific theorizing.

In order to facilitate an understanding of the way religious commitment and faith influences scientific theorizing, I shall make use of an understanding of religion that deviates from the conventional understanding in which it designates a restricted sacred area of human life in distinction from the profane or the secular. The claim in this chapter is that there is a clearly discernable influence and presence of religious convictions in all scientific theorizing. This statement requires implementing the broader understanding of the nature of religion and the unpacking of the role of the fiduciary dimension in cognition by addressing three issues: The nature of religion; the fiduciary, confessional or certitudinal dimension of human embodiment, and the interaction between religious beliefs, metaphysics and science. The role that the fiduciary dimension plays as a vehicle for religious convictions in scientific theories, is developed in the next chapter.

Religion in the Broad and Narrow Sense

The practice of religion in the narrow sense of the word has to do with creeds, dogmas, liturgical practices and beliefs that relate to an individual or community's understanding of the worship of that which is regarded as Divine. In the broader sense of the word, the notion of religion employed in this chapter pertains to a total all encompassing commitment to that which is regarded as ultimate. This commitment is an all-encompassing relationship, which permeates all facets of human life and comes to a differentiated expression in every single dimension of human life. Clouser (1991:21–23) defines religious belief as '[…] a belief in something(s) or other as divine' or '[...] a belief concerning how humans come to stand in proper relation to the divine'. The 'divine' is characterized here as that which can exist independently of anything else. In his revised work Clouser states that the definition of religion needs to be expanded as follows:

> (1) A belief is a religious belief provided that:
> (2) It is a belief in something as divine per se no matter how that is further described, or
> (3) it is a belief about how the non-divine depends upon the divine per se or
> (4) it is a belief about how humans come to stand in proper relation to the divine per se,
> where the essential core of divinity per se is to have the status of unconditionally non-dependent reality (2005:24).

This is obviously a redefinition of the standard notion of 'religion'. One needs to differentiate between religion in the narrower sense of the word i.e. certitudinal or confessional acts pertaining to worship and cultic acts and religion in the wider sense of the word, pertaining to the commitment of the totality of human life to that which is regarded as ultimate. This wider definition of religion makes it possible to distinguish between typically confessional or certitudinally qualified acts of worship, prayer, liturgy, etc., but also accommodates the fact that in all aspects of human life there are analogous, typically qualified, expressions of a certitudinal or confessional nature. Examples are the 'confessional element' present in the policy statement of any political movement or the 'confessional element' in the scientific statement of the *Wissenschaftliche Weltauffassung. Der Wiener Kreis* (Neurath, 1973). In the latter two examples, the certitudinal elements function in an analogical way, i.e. analogous to religious certitude. They are not 'confessional' in the original sense of a confession of faith, but are analogous to it. When religion is understood in these senses, it changes the way one formulates and resolves the issue of the influence or presence of religion in science. It is to this issue that we now turn.

Science attempts to describe, explain or interpret different dimensions of reality and in this process utilizes theoretical constructs and concepts to approximate the patterns or order which characterize the explanandum. Scientific theorizing is hypothetico-structural explanation of the explanandum. The expression of religious commitment in the area of scientific endeavour differs from its expression in other areas of human life. One could say religious influence in science is garbed in the robes of the structure of science and when one is pursuing the interrelationship between science and religion or the influence of religious convictions in science, this definition of religion would direct the search for convictions that qualify as religious in a different direction than would the standard definition of religion. One would pursue the certitudinal, confessional or fiduciary convictions that are analogous to religious convictions and which influence science intrinsically and have epistemic import. Such a pursuit rests on a number of assumptions:

- That all of human life is religious in the sense of relating to the Divine or to what is regarded as the Divine.
- That human embodiment exhibits a fiduciary dimension[1], which comes to expression in all human experience, activities and events.
- That this fiduciary, confessional or certitudinal dimension is characteristic of religion in the broader sense of the word.
- That this fiduciary dimension is also characterized by the principle of aspectual universality, as are all other aspects or facets of human existence and reality.
- That the structure of theorizing incorporates a fiduciary dimension.

In a different context, Olthuis (1987:24) argues that we are equipped with 'our believing-entrusting-faith way of being in the world [...]'. He gives this way of being in the world the technical name certitudinal (1987; 1985:53–164.) When Olthius discusses the term 'certitudinal' he has in mind belief/faith in that which is ultimate or that which is regarded as ultimate, i.e. what generally

1 Steiner (1989) recognizes a similar aspect of reality which he calls 'trust' (1989:89, 91, 148). He says: 'There would be no history as we know it, no religion, metaphysics, politics or aesthetics as we have lived them, without an initial act of trust, of confiding, more fundamental, more axiomatic by far than any 'social contract' or covenant with the postulate of the divine'. Clouser (2005:246) uses the term 'fiduciary' to refer to '[...]the varying levels of reliability or trustworthiness a thing or person may have'.

would be regarded as religious faith. But structurally this same human propensity forms an integral aspect of all human action and cognition. Reliance, acceptance, surrender (Hart, 1984:183) occur in all human relationships. People accept themselves; rely on one another and the world around them. Some examples are: surrendering to the inevitable, accepting reports on face value, believing in human rights, committing to the terms of a contract, committing to a relationship, trusting a bridge – these are all instances of the human fiduciary or certitudinal image schematic structure and domain of experience. I will now develop some arguments for the recognition of a fiduciary dimension of human embodiment and then show how this insight can be used in the understanding of the presence and role of fiduciary control beliefs in scientific theorizing. For this we need to return to the notions of embodiment and conceptual metaphor which were addressed earlier.

The Omnipresent Fiduciary Dimension

Johnson (1989:109–118 and 1991:3–18) states that embodiment is the locus of our experience and specifies embodiment more precisely as the patterns that emerge through the sensori-motor activities of human beings (1991:8). This has to do with spatial and temporal orientation, bodily movement, perceptual focus and manipulation of objects to accomplish certain purposes. In the use of these sensori-motor skills, image schematic structures emerge and develop which provide the form, order and relative determinateness in human experience of the world (Johnson, 1991:10). These image-schematic structures are also at work in the more abstract realms of understanding and reasoning Johnson (1991:12). All of this Johnson claims is knowing. The image-schematic patterns are what Johnson calls 'imaginative structures' that are the imaginative contours of our experiential interactions that are employed in order to have ordered, recognizable representations and experience[2]. The human ability to believe, to trust is one of these image schematic structures.

The fiduciary 'moment' is an integral element of human embodiment, experience and the act of cognition but also of the multifaceted world, which the knower attempts to know. Reality and cognition exhibit the same multi-faceted stratification. These aspects of reality are integrally related, irreducible to one another and yet contain implicit references to other domains of experience. These potential possibilities are not necessarily made patent. They are opened up in the interactive embodied relationship of actors and the world around them. The original basic equivocal concept of certitude acquires a multiplicity of meanings. The fiduciary moment, which expresses the human ability to believe (fides), to trust, to be certain, is an integral element of both a multi-faceted world and its potential to form the basis of a multiplicity of ways of knowing this world. In all human activity there is such a fiduciary element present e.g. a tight rope walker trusts his abilities, or the security of the rope or the safety net, the client trusts the banker, the marriage partner commits himself in troth to his partner, etc. Examples of such analogical relationships between the fiduciary aspect and other aspects are: social credit, legal trust, moral troth, psychological assurance, promise (a promissory note), confidence, credibility (financial or otherwise), etc. This state of affairs points to the fact that the multi-faceted and multi-dimensional world exhibits both irreducible aspects and an integral internal coherence of these aspects.

2 Cf. Rohrer's (2001) summary of the tenfold meanings of embodiment.

In the terminology of recent Cognitive Semantics one could state that these examples point to the fiduciary dimension of reality as a conceptual metaphor. This still leaves an important issue unresolved: Are these conceptual metaphors merely bodily based phenomena or is there possibly a more fundamental ontological issue at stake in the uncovering of these apparently subjective states of affairs.

The Fiduciary Rootedness of Cognition

This tacit fiduciary dimension is present in all aspects of reality and therefore also in cognitive acts. It characterizes the commitment an intellectual community shares when viewing the similarities and dissimilarities of the world in a certain way. One could perhaps even speak of a 'dogmatic' dimension in this respect. Somewhere Kuhn states that his initial notion of the role of 'dogma' in scientific research is one that he actually does not pursue any further. But it is exactly the presence of such a 'fiduciary' element which is present in all human acts that points to a domain of experience which is as universal as spatial, numerical, kinematic, physical, biotic, psychic, linguistic, social and cultural domains. It is also this religious-like, dogmatic commitment to a certain framework that determines how the world is perceived and conceived both in everyday life and in science. The fiduciary dimension of cognition is an image-schematic pattern characteristic of all human existence, interpersonal relationships and their relationships to reality.

In the context of discussing personal knowledge Polanyi refers to a related notion. He speaks of belief as the source of all knowledge and continues: 'No intelligence, however critical or original, can operate outside such a fiduciary framework' (1974:266). He also speaks of the 'fiduciary rootedness of all rationality' (1974:297). Whether such a dimension is actually constitutive of human experience and as such a basic experiential domain can also be demonstrated by showing the negative consequences of a lack or absence of trust, certitude, or belief in human relationships. Polanyi states that even a '[...] programme of comprehensive doubt (when someone claims to doubt all)[...]' (1974:297), collapses and reveals by its failure the fiduciary rootedness of all reality. Polanyi argues that any range of conscious awareness always still presupposes the uncritical acceptance of some beliefs.

Polanyi's view of the fiduciary nature of cognition does not fully coincide with the view developed here, but provides an important argument to show that there actually is such an image schematic structure present in our experience of reality. The fiduciary element in cognition in general (appealing to 'fides') needs to be recognized in notions such as 'con-fidence', conviction, credit, etc. In all these acts an element of belief is present. Not only is this a structural human ability but when human beings interact with the world around them or have experiences in the world, these objects can become the 'objects' of their certainty or they can 'rely'/trust on them. Things in the world, other people and relationships have a dormant fiduciary dimension that can be disclosed.

The Fiduciary, the Spiritual and Religion

In the fiduciary function which is present in all human acts and interactive relationships between actors and the world, all other 'radial' aspects of reality and of human experience come to expression in an analogical fashion. The existence of and specific understanding of the fiduciary aspect of reality, human existence and cognition is not addressed in the work of Lakoff and Johnson, but they do relate metaphor to embodied spirituality. Their notion of spirituality approximates the use of the term religion by Clouser (1991; 2005). The significance of this reformulation of the notion of religion is that religion is now not necessarily related to God/gods, worship, liturgy, moral and ethical practices, but to that which is regarded as divine and therefore non-dependent. Religion is a commitment to that which is regarded as ultimate, or metaphysically non-dependent or the final ground of all things, which require no further grounding. One's ability to trust is instrumental in this religious commitment, but is not synonymous with it. Distinguishing this notion of religion from the fiduciary moment characteristic of all human experience, knowledge and reality facilitates a clearer understanding of the presence and influence of 'religious' convictions as ultimate convictions and the (certitudinal, fiduciary) convictions, which mediate this ultimate conviction in scientific theories. This presupposes two assumptions. The first is that all of reality is meaning (not has meaning) and the second is the acceptance that all human experience is rooted in a deep and profound quest for (religious) meaning. Metaphorical mappings and analogical meaning transfer are some of the pivotal mechanisms which mediate meaning disclosure in a great variety of areas. One could speak of the mediating and hermeneutical function of metaphor. All acts, events, facts, relationships are involved in this quest and moreover exhibit a certitudinal or fiduciary aspect. Embodiment theories have provided important clues to attempt an answer to this question and have also opened avenues to recognize the role of fiduciary elements in cognition. We shall now attempt to trace the role of the fiduciary element in scientific theorizing.

Approaches which argue that religious presuppositions and religious commitment influence scientific theories often neglect to demonstrate how this is the case. Metaphorical hermeneutics attempts to demonstrate this required how-relationship. Central to this methodology is the thesis that the presence and influence of religious notions become apparent in the way theoretical concepts are formulated and function within theoretical frameworks.

The methodology questions two positions: The first is the radical positivist distinction between theory and observation in which meaning *invariance* of concepts are postulated. The second position is the radical theory-leadenness of observation thesis in which conceptual meaning-variance ushers in potential relativism. Both these positions cannot avoid the pitfalls of a reductionist theory of meaning. Instead, this approach assumes a network theory of meaning[3] (suggested by Nersessian, 1987 in conjunction with Quine's theory).

The network theory of meaning takes into account the presence of both epistemic and so-called non-epistemic factors (inter alia the dimension of discovery and the representational resources that are part of the scientist's culture) in the process of scientific discovery and theoretical concept formation. A network theory of meaning rejects both the double language thesis with its radical distinction between literal and metaphorical language and the radical distinction between observation and theory in science. It argues

3 Cf. Leplin (1988)-Nersessian (1991b:678-86) discussion, which addresses the feasibility of such a theory of meaning; requires rejection of the radical distinction between literal and metaphorical language (Hesse, 1970, 1983; 1985/6; Botha, 1986b) and also implies that the distinction between observation and theory in science is relative and contextually determined.

that these distinctions are relative and contextually determined. Scientific concepts are embedded in a network of beliefs in which metaphysical beliefs are as constitutive as any other type of belief present in the network. These scientific concepts are regulated, determined and conditioned by the choice of ontological strata and conceptual schemata which limit their parameters of meaning. One such schema is the fiduciary dimension of reality which is constitutive of cognition in general and also of theorizing.

The Fiduciary Element in Theorizing

The expression of this fiduciary dimension is to be found in the philosophical assumptions undergirding a theory. An interesting example in which the reliance on and presence of philosophical assumptions in theorizing is acknowledged is General Systems Theory, a popular and prevalent approach in Biology, Sociology, Social Work, Political Theory and various other disciplines. Weltman (1973:14) recognizes this and points to an even more fundamental characteristic of these philosophical assumptions. He calls the underlying structure of reality uncovered by Systems Theory 'the ultimate reality'; a reality which cannot be proved or disproved, but can only be stated. He argues: 'It is basically a metaphysical scheme which seeks to find in reality an ultimate orderliness above and beyond the ken of normal scientific methods of verification' (1973:14).

These metaphysical schemes harbour the ultimate commitments of theories, just as human stories are fundamentally the stories of human lives being lived within the parameters set by deep personal religious commitments. When Laszlo, one of the expositors of Systems Theory, extols the virtues of Systems he exhorts us to revere 'natural systems'. He sees man and nature as an '[...] embracing network of dynamic, self-regulating and self-creative processes'. He continues: 'To know this is to admire the matrix out of which we arose, and to want to preserve what we are. We do not need elaborate fantasies or tales of legendary creation' (1972:289). This comment tells its own story. These are philosophical assumptions that undergird his systems view of the world, philosophical assumptions that fulfil a role very similar to that which creation stories would within the life of a religious community or individual believer. The metaphor of man and nature as 'self-regulating and self-creative' in this theory functions analogous to a religious belief. It is an ultimate belief expressed in a metaphor and rooted in a deeper basic systems metaphor. The systems metaphor functions as root-metaphor in this theory. These root-metaphors do not only function within scientific theories but are part and parcel of the wider culture and society within which science functions.

Rethinking Root-metaphors: Re-Enchanting a Disenchanted World

Exploration of the rhetoric concerning the events and images of 9/11 show that a vast diversity of metaphors are chosen to depict, describe or explain the nature of the events. They range from depiction of the events as war, crime, violence and terrorism to sophisticated theoretical analyses from the vantage point of psychoanalysis to systems theory. A complex and multifaceted event such as this is obviously open to manifold vantage points and interpretations. The preference for and choice of any one specific metaphoric description is not a simple and innocent matter – it can have serious repercussions. But, it is not only the metaphors utilized in the description or explanation of the events that are of interest here. It is the challenge to show the deeper ideological nature of the motivation at work in the underlying root-metaphor which concerns us. It is the need to develop or discover a methodological tool that can assist in the diagnosis of the 'ideological tilt' that some metaphors acquire which constitutes the main question of the next chapter.

Twenty years ago I attempted a similar analysis of the fundamental religious motives at work in the (apartheid) South African political situation where an analogous 'terrorist threat' existed to the political establishment of the time (Botha, 1983). Historical hindsight has shown how deeply embedded the events of the time were in 'civil religious' ideologies – operative in both White Afrikaner nationalism and Black African nationalism. The need was then as it is now, to discern a methodology that will help to distinguish 'ideology' from worldview, myth and religion. Phrased differently: We are in need of a methodology that indicates when metaphorical symbolization is authentic or inauthentic (Winter, 1981). For this purpose we look at the mediating role of metaphor in the relationship between religion, metaphysics and science.

VI Possible Worlds: The Mediating Role of Metaphor Between Religion, Metaphysics and Science.

When do Standard Cultural Allegories or Licensing Stories Become Reified Myths?

The powerful images of the events of 9/11 have made an indelible impression on the world psyche. It has given rise to a pervasive rhetoric in practically all fields attempting to explain, interpret and understand the underlying causes and world changing consequences of the events. In a postmodern and secular world it has led to refocusing on the religious fervour and ideals at work in established religions and in movements that are ostensibly devoid of all religious motivation, such as expansive globalization and American free enterprise capitalism. Although ostensibly religiously neutral, there are ideological and 'religious' notions at work in the conflict of worldviews represented by these movements. This becomes apparent when root-metaphor analysis reveals the underlying 'war of worlds' at work in the foundational symbolizations of the world (Winter, 1981). These foundational symbolizations function as fundamental building blocks of the cultural cosmologies of society (Harrington, 1995:360). Not only do religious convictions shape the world but they also shape the choices of metaphors by scientists, educators and politicians. These ostensibly secular choices are not random and neutral but laden with root-metaphorical notions and are as religious as those made by their overtly religious counterparts. The choices are embedded in theoretical and metaphysical frameworks that dramatically alter the perception of phenomena and the behaviour and actions of groups. Recognition of the presence and influence of root-metaphors of an ideological nature will contribute to the disenchantment with their allure. This chapter builds on the methodology of metaphorical hermeneutics and its assumed ontological stratification in order to identify the presence of root-metaphors and unmask the reductionism and hypostatization which often characterizes their function in science and society. The crucial question posed here is: What role is played by root-metaphors or worldviews and their control beliefs in the process of science becoming ideology?

Both ordinary experience and sophisticated scientific experience require criteria that would make it possible to determine semantic conflation (Spragens, 1973:41). In scientific theorizing this is important in order to avoid becoming the victims of a situation where, 'metaphors become myths' (Turbayne, 1970:28). From very ordinary, everyday mistakes to the identification of ideology, human beings make judgements based on categories and classifications that are not merely conventional and sociologically determined. These judgements embody views of reality and at times distorted views of reality. The important question is what the criteria are on which the diagnosis of 'distortion' is based. This obviously has implications for the diagnosis of ideology. Laszlos' (1972) version of Systems Theory is a good example of the way in which a perfectly legitimate metaphor and model is loaded with philosophical assumptions that not only function as religious control beliefs, but also veritably become religion or ideology. So the question is: At what point does the legitimate use of metaphor exceed its boundaries?

There is agreement that the moment the limited perspective, provided by the metaphor, is assumed to be able to explain the full reality it is dealing with, it has moved in the direction of derailment. Most probably this was meant by those early analysts who claimed that metaphor became myth once it was applied literally (cf. Turbayne, 1970). The problem of derailment arises once the basic metaphor is stripped of its contextual limitations and extrapolated far outside of the limited boundaries within which it can legitimately operate. Once the root-metaphor in a theory exceeds its boundaries as legitimate control belief it starts functioning in an ideological fashion. Such an argument rests on the recognition of limits, boundaries, order and contexts – the ontological issue germane to the whole analysis of metaphorical hypertrophy. Although (Von Bertalanffy, 1952:1972), the founder of General Systems Theory was conscious of the limitations of his own theory, Laszlo independently developed a theory with far more pretentious expectations. According to Laszlo 'We are part of the tremendous balanced hierarchy which distributes roles and values to all things, from atoms to international federations' (1972:289). The ontological holism of Laszlo (cf. Laszlo, 1972:291) is justifiably qualified as 'A Study in Metaphoric Hypertrophy', the subtitle of Weltman's work (1973:82). The limits and boundaries of metaphor discloses the role that the ontological furniture of the world plays in the constraints and possibilities of meaning disclosure that can be mediated by a metaphor. In this process of mediation basic – or root– metaphors play a significant role.

A helpful distinction between the different types of metaphors and their divergent functions is one that differentiates between metaphor in language in general, in model building in science and the more basic 'root-metaphor' function of metaphors that operate in myths, religion, world views and in science. A worldview can be described as an encompassing framework of basic convictions concerning all things in reality and can form the basis of various theoretical systems (Wolters, 1985:2). Pepper's (1942; 1970; 1982) notion of a root-metaphor or world hypothesis is probably the most well known in the literature today. He emphasizes the fact that root-metaphors are often generated from everyday experience and that some insight from this experience is used in an analogical sense and extrapolated to all other areas of reality in order to describe or explain reality. A root-metaphor aims at bringing some large area of reality into perspective and differs from paradigms or metaphors of restricted scope which he calls cognitive metaphors of restricted scope (Pepper, 1982:202).

Brown (1977) more or less identifies a root-metaphor with paradigms (Kuhn) and argues that it is often also used in the sense of a philosophy, metaphysics or worldviews of a certain domain. Brown defines a 'root-metaphor' as follows:

> Root-metaphors are those sets of assumptions, usually implicit, about what sorts of things make up the world, how they act, how they hang together and, usually by implication, how they may be known. As such root-metaphors constitute the ultimate presuppositions or frame of reference for discourse on the world or any domain within it (1977:125).

The root-metaphors that are fundamental to scientific theorizing are usually none other than regular ordinary language metaphors or scientifically developed and articulated notions of these metaphors. Philosophical presuppositions are the articulations of these religious root-metaphors which determine the nature of and provide the parameters within which the scientific control beliefs of theories (Wolterstorff, 1984) function. The basic metaphors or root-metaphors provide the scientific community with a set of control beliefs and paradigmatic propositions which regulate the research process. Only when the role of root-metaphors and world views are taken into consideration in the analysis of the structure of science, does it become possible to give an adequate account of the possibilities for and nature of ideological derailment of scientific theories. This argument requires a definition of the nature of ideology that can accommodate the controlling, regulating and directive influence of fiduciary beliefs in the structure of theoretical or scientific belief systems. It must be able to also accommodate the negative derailing and distorting role that metaphors can play. If metaphors are as fundamental to cognition and human interpretation of the world as is currently being claimed in a variety of disciplines, their potential misguiding and misleading role in the distortion of reality also needs to be addressed. What should be regarded as an 'authentic' or 'inauthentic' metaphorical symbolization?

Contrary to definitions of ideology which see it primarily as a system of thought of a socio-political nature the term 'ideology' is here defined closer to Dirven's notion of a '[...] social myth by which a community lives [...]' (1994:152). But the notion here is even stronger and depicts a metaphorical framework which is viewed as negative and detrimental to a true understanding of reality. Ideology is being used in the derogatory sense of the word indicating a derailed and distorted view of events and reality.

Scientific worldviews and broader cultural worldviews often develop based on the same basic metaphor or basic analogy about the structure of the world or experience. However, they differentiate respectively as pre-theoretical or unsystematic and theoretical or systematic articulations of these worldviews. The basic image or picture of reality (scientific world 'view'/*Weltbild*) provides the categorial system in terms of which the scientific discipline or theory's 'world' is seen. It provides an account of the purported structure of the explanandum and all the entities, processes, relationships and laws claimed to be present in it. It conditions the metaphors and analogies which are regarded as acceptable in order to approximate and articulate the structure of the entities, laws and processes in the domain under discussion. Such a basic view, provides both the basic and conveyance metaphors (MacCormac, 1985:46–48) which structure the generation of hypotheses and the formation of concepts in theorizing. For this reason metaphor is such an appropriate site for studying the broader cultural dimensions of science. A good example is provided by Cantor (1987:124–143) who shows how the fluid, sand and ballistic metaphors and the language of optics were impregnated by theological discourse and theological values. This also holds true for language in general which is often impregnated with the changing ideologies of the wider cultural milieu (Hesse, 1980:252). To accommodate this type of metaphor the term 'basic metaphor' is introduced.

The term basic metaphor is closer to the sense of Kuhn's (1977) global notion of a disciplinary matrix and is often substituted for the term root-metaphor. The latter is often used in both a wider (Pepper, 1982; 1942) or a more restricted (McMullin, 1978; Cantor, 1987) sense. A basic metaphor provides a conceptual tool for the configuration of domain assumptions based on a root-metaphor, and makes it possible to distinguish it from the more specific metaphorically constituted (Soskice, 1985:102) theory terms. The latter type of metaphors could also be called conveyance metaphors. Conveyance metaphors are the articulation of basic metaphors. They are more restricted in scope than basic metaphors and often grow out of basic metaphors. The seventeenth century mechanization of the world picture (Dijksterhuis, 1969) is an example of such a pervasive metaphor which gives rise to mechanistic domain assumptions in the work of a variety of physicists and philosophers.[1]

This basic metaphor is often, but not necessarily, of religious origin but functions in a way analogous to religion. It dictates and prescribes the choice of analogical emphasis in the scientific metaphor[2]. This emphasis functions as control belief in the configuration and cluster of beliefs that make up a theory. Contrary to general opinion, religious factors are not per definition external and non-epistemic when it comes to theorizing. Because fiduciary traits are as intrinsically embodied in cognition as are other aspects, these religious factors can function externally, internally or constitutively in scientific theorizing.

There is a relationship between a basic metaphor which provides the categories in terms of which the world is seen and the specific type of metaphorical control beliefs that function in a fiduciary fashion in ordinary and theoretical concept formation. These metaphorical control beliefs are central to the understanding of embodied epistemic faith in the process of cognition. Such a basic metaphor or root-metaphor conditions the metaphors and analogies regarded as acceptable for the articulation of the structure of the explanandum, provides a way of speaking about and seeing phenomena and structures the generation of hypotheses and formation of concepts in theorizing. In the metaphor a set of fiduciary beliefs – so-called 'core beliefs' – function in a 'religious'

1 Cf. Chapter VIII on Religion, Metaphysics and Science in Michael Faraday's work.

2 Kuhn points to this phenomenon in his exposition of the nature of a paradigm (2000:533ff.)

fashion. This refers to the way these beliefs are held and not to their content. The perspective on any phenomenon changes when the set of beliefs that acquire primacy and guide the focus of the metaphor, changes.

Ideologies: Metaphors Gone Awry?

What is at work in ideology is the presence of dysfunctional controlling analogical moments in the use of the metaphor. In the case of an ideology the normal function of fiduciary elements in the conceptual belief structure of theories have become malignant. These fiduciary elements in worldviews and social philosophies are also akin to those found in the general conceptual structure of all belief systems. Their presence does not make them prone to ideological derailment. But, when the metaphor at the base of the conceptual structure somehow truncates, reduces or hypostasizes an aspect of reality it inevitably leads to ideology. The mere presence of fiduciary elements in belief structures as such is not the decisive element in the development of ideological interpretations and constructs, but when such fiduciary elements are elevated to the level of the only and primary mode of understanding or explanation, the seed for metaphorical hypertrophy has been sown. They can then function as ideology both in theoretical and pre-theoretical garb.
Harrington's (1995:360) analysis of the role of 'holism' in the Third Reich clearly shows that both in scientific theories and in the National Socialist culture an organic holistic metaphor was at work – one in which the biological metaphor of an organism formed a 'cultural cosmology' or 'meaning world' which dictated human action in opposition to what was regarded as atomistic and mecha-

nistic forms of thought (1995:366). Her argument shows that a cultural cosmology which originates from a scientific and theoretical understanding of reality can seep into the surrounding culture and acquire the role of an ideology. The puzzling question is how one determines the legitimate role, limits, and boundaries of the metaphors at the root of the cultural cosmology. I suggest that Lakoff and Johnson's (1981) construct of a universal metaphorical conceptual structure needs to be augmented and modified to accommodate the recognition of the analogical nature of human embodiment and the analogical stratification of reality. This analogical stratification of reality, embodiment and human experience is prior to human thought, concept formation and language and comes to expression in metaphor mediated experience. When these characteristics of the world and of human experience are articulated in worldviews and theoretical paradigms based on root-metaphors, they inevitably reveal implicit fiduciary elements, i.e. decisive and regulating beliefs that dictate how the world is to be seen and acted in. Ideology can only be counteracted when the intrinsic limitations of such metaphorical interpretations are acknowledged. How is this done? How does one escape the trap of the 'faulty logic of one's own metaphors [...]' (Harrington, 1995:374)? It is obvious that this can only be a futile 'bootstrap operation' when no vantage point outside of the metaphorical system is found. Does such a vantage point exist? An important step in the process of developing an answer to this question is recognizing the deeply religious nature of views of reality and specifically of those views that have been declared absolute.

Ideology reveals itself by the implicit or explicit claim that it represents the only legitimate way of looking at the world and that it therefore functions as the only salvific way to achieve a world according to its precepts. As important a step is judging the consequences of such views. When root-metaphors in science or in society have in their wake the miscarriage of justice and the eclipse of mercy or the rape of the earth, they are harbingers of evil and require being exposed. History exposed South African 'apartheid' policies as such – hopefully philosophers will have the courage to expose the 9/11 rhetoric in a similar fashion. An important instrument in this expose will be a return to the recognition of the ontological basis of metaphorical language and the limits and boundaries imposed by them.

Eubanks' analysis of the role of 'ideological freight' in metaphorical meaning is helpful in this respect. He claims: 'No metaphor comes without ideological freight' (1999:419–442). He speaks of the 'cultural motivation of metaphoric mappings' (1999:421). Aptness of metaphors and metaphorical mappings proved to be determined by 'licensing stories' (truism, personal anecdotes, recounted stories reflecting their ideologies). Eubanks argues that his re-examination of metaphor reveals complexly operating rhetorical patterns, that these patterns help to constitute conceptual metaphors and that the development of a richer account of metaphor as a cultural phenomenon is possible if the patterned relationship between metaphor and other discursive forms like 'licensing stories' are considered (1999:420). His research showed that a 'licensing story' expressed the discussant's view of '[...] how the world does work and how the world should work' (1999:424, 426). These licensing stories are often embedded in or form the basis of a 'standard cultural allegory' (ibid. 1999:429). Eubanks writes:

> Licensing stories are narratively structured representations of an individual's ideologically inflected construal of the world. Metaphor aptness – which is to say, the aptness of possible mapping – depends crucially upon this construal. (1999:437).

Eubanks argues that conceptual metaphor is inflected at a minimum by politics, philosophy, social attitudes and individual construals of the world (1999:420,437) specifically. He argues: 'Conceptual metaphors are inseparable from the circumstances in which they are uttered and thus they are always inflected by discursive conventions and ideological commitments' (1999:422). These standard cultural allegories or root-metaphors function as the licensing stories for metaphor mappings. In the context of theories this claim requires further specification.

Since the original implementation of the notion of a 'root-metaphor' by Pepper (1942) the notion has been appropriated for a variety of uses. The argument here is that a root-metaphor or 'licensing story' contains a number of absolute presuppositions concerning the nature of reality. They are protected from possible falsification and function analogous to dogmatic faith statements in religious systems and religious experience (Miller, 1969).

A somewhat loose reciprocal relationship can be construed between a broader worldview or myth and the metaphors legitimized by such a myth or worldview. When such a worldview acquires the status of a root-metaphor or functions as a paradigm in theorizing its constraining and limiting function with respect to the permissible choice of metaphors, becomes more pronounced. In this process some or other element within the radial scope of multivalent meanings within the metaphor gains primacy and acquires a controlling function. This becomes apparent in the meaning of the analogical concepts utilized in theorizing.

Metaphors that function in poetic or literary discourse or in everyday language refer to concrete entities and can often be replaced by other, striking metaphors without loss of meaning. The type of metaphors utilized in scientific theorizing and argument make use of ontic analogies. No theory can do without metaphorical references of this nature. Most probably this is what Ortony

(1975) refers to as *necessary* metaphors. Metaphors often function as the vehicles of the fundamental analogical concepts of the specific sciences; i.e. those fundamental concepts without which the practitioners of a discipline cannot theorize at all. These fundamental analogical concepts refer to the structural joints of reality. To the extent that theories developed on the basis of these concepts approximate these 'joints', the metaphors mediating this approximation are scientifically apt, fitting or suggestive. This becomes apparent when a revolutionary scientific innovation introduces a new root-metaphor into the scientific terminology of a discipline. How ideological and socio-historical factors condition, constrain and/or determine what is regarded as apt or fitting meanings and appropriate meaning transfer, ties into the role of standard cultural allegories and licensing stories in the use and abuse of metaphor. The theory-constitutive role of metaphors in science provide an important clue in this regard.

VII Theory-constitutive Metaphors in Science

Tracing the Constitutive Presence of Religious Convictions in Science

The nature of the influence of religion in science varies. Brooke (1991:16–51) lists the following possible relationships that can be discerned in the history of the relationship between faith and science: Religious convictions as presuppositions of science; as sanction for science; as motive for science. Religious convictions in a selective role; as regulative principles; or in a constitutive role in science. Additional to these relationships, religion can also function as the source or origin of creative and innovative ideas for theorizing. Religious considerations, notions, ideas and concepts can externally, contextually condition theories but can also penetrate, pervade or permeate theories internally. It is especially the latter type of function that we have in mind in this chapter and it correlates primarily with Brooke's last two types of functions, i.e. religious convictions as regulative principles and their constitutive role in theorizing. In this chapter the emphasis will be on the penetrating, pervading or permeating role of religious convictions in scientific theories. It is their internal or constitutive role that I have in mind when we speak of the role of religious convictions in scientific theories. In the latter sense the term refers to 'the manner in which these notions function,' or the way in which these beliefs are held. This is the pervasive and permeating influence of a set of basic or core beliefs, assumptions or presuppositions which are given privileged and priority status in the process of theorizing.

It is especially the role of those control beliefs which function in a religious fashion, that I have in mind. Internal scientific control beliefs need not necessarily be religious in either origin or content in order to function religiously (i.e., analogous to religious belief) within the belief structure of scientific cognition. Any belief within the belief structure of scientific knowledge can be elevated to the status of a control belief. When such basic beliefs, assumptions or presuppositions of science are held in such a manner that they are protected against all possible falsification they could be characterized as scientific beliefs acting in a religious manner. Such beliefs are often called a certain type of metaphysical belief or absolute or paradigmatic presuppositions (Collingwood, 1940; Burtt, 1954; cf. Luckman, 1986; Brown H., 1975).

Religiously determined basic beliefs in science have to be sought within the scope of philosophical assumptions and metaphysical beliefs which guide or control research and theorizing. Not all metaphysical beliefs and philosophical assumptions necessarily have this guiding/controlling function. Theory-constitutive scientific metaphors (i.e. metaphors which propose models) often reflect and are permeated by deeper basic metaphorical notions some of which are intrinsically religious in nature and determine which analogical moments of a theory-constitutive metaphor regulate, control, steer and guide the process of theorizing. These fiduciary factors enter theorizing via the focal analogical moments that acquire a controlling position within a theory. They are not 'non-epistemic' but constitute an integral dimension of the process of cognition. The meaning of concepts utilized in the linguistic projections of theory-constitutive scientific metaphors are mediated by such focal analogical moments. These in turn reveal the basic metaphor on which the theory is predicated. Core or control beliefs in science, function legitimately when they steer, guide or control theorizing with respect to a limited facet of reality and illegitimately when they either implicitly or explicitly function as explanation of all of reality. This can often be recognized in the phenomenon of *metaphorical hypertrophy* (Weltman, 1973). This phenomenon will be demonstrated when the work of Michael Faraday is introduced.

Knowledge acquisition is also inextricably related to the structure of the world it is attempting to explain. The analogical moments within the radial spectrum of meaning of a conceptual metaphor form an integral part of the metaphor. This is similar to Lakoff and Johnson's argument that a radial category of concepts has '[...] a central member and extensions, many of which are metaphorical' (1999:499). In the metaphor the cross-domain mapping establishes a focus or screen which opens up some dimension or aspect of reality. This correlates with the intrinsically 'metaphorical' structure of reality.

Metaphor and Scientific 'World Views'

One of the objectives of this work is the exploration of the significance and impact of embodiment theories of metaphor on some traditional formulations of the relationship between science and religion. Theories about the pivotal role of metaphor in language, cognition, creativity, meaning and a host of other areas abound. Two issues central to all of these accounts are the mediating and hermeneutical functions of metaphor and the way metaphorical meaning is conditioned. The fiduciary dimension or aspect – that which pertains to trust plays a pivotal role in the cross domain mapping which a metaphor creates. This element mediates the controlling or guiding analogical moment of a basic or root-metaphor. Religious control beliefs (Wolterstorff, 1984) or fiduciary beliefs (Polanyi, 1974) refer to the way these beliefs are held and not to their content. In the use of the systems metaphor it would in principle be possible to emphasize either the mechanical, organic or cybernetic dimension.

The cross domain mapping that occurs in each case discloses a different field of meaning. In the configuration of analogical elements present in such images one analogical element acquires primacy and guides the focus and perspective of a theory. These analogical configurations are often deeply embedded in the dominant worldview and root-metaphor of an era[1]. The metaphor provides an account of the purported structure of the domain of discourse and its relationships to other domains. It conditions the metaphors and analogies, which are regarded as acceptable in order to approximate and articulate the structure of the entities, laws and processes under discussion. Such a basic view provides a way of speaking about phenomena which structure the generation of hypotheses and the formation of theoretical concepts. Theory-constitutive scientific metaphors are metaphors which propose models and often reflect deeper basic metaphorical notions that function in an intrinsically religious way. These basic metaphorical notions determine which analogical moments of a theory-constitutive metaphor regulate, control, steer and guide the process of theorizing. It is the relationship between such a 'basic metaphor' and the specific type of control beliefs that function in a religious fashion in concept formation and theorizing that is of interest here. In science these basic metaphors are embedded in the disciplinary matrices or paradigms that characterize the discipline.

1 Cf. the work of Anne Harrington 1991; 1995; 1996 with respect to the influence of the organismic and mechanistic world views in German science and society.

Paradigmatic Presuppositions

Scientific theorizing presupposes a commitment to a set of philosophical assumptions and presuppositions, some of which are held in much the same manner as religious convictions. These presuppositions, which constitute part of the scientific worldview, are held in such a way that they guide and control all research which fall within their scope and are carefully protected against possible falsification. When metaphors are used in a presuppositional sense as root-metaphors on which scientific theories are built, they determine the ultimate presuppositions or frames of reference of scientific worldviews and are often the source of the control beliefs which function in a confessional, certitudinal and fiduciary fashion. Carl Rogers' (1981:ix, x) conception of the formative directional tendency operative in reality constitutes an example of a presupposition which functions as a control belief in his theory of man and the universe. It is an instance of the more encompassing root-metaphor of holism.

Both science and religion harbour a certitudinal or fiduciary dimension which plays a pivotal role in its universe of discourse. These are the level of ultimate or absolute presuppositions concerning the nature of reality (Collingwood, 1940). Brown calls them paradigmatic propositions and states that they '[...] play a fundamental role in determining what science is in any era, how research is to proceed and how our observations are to be interpreted' (1975:86,88). He argues: 'They are fundamental presuppositions which organise and guide scientific research and without such presuppositions no coherent research is possible' (Brown, H. 1975:87; cf. also Brown, R. 1977; Brown J. 1979). Wolterstorff speaks of a similar phenomenon and calls it control beliefs (1984:69) He says they fulfill a regulative function in scientific theorizing. It is the function of these beliefs and of absolute presuppositions which are intrinsically religious (fiduciary) in nature, i.e. they show marked family resemblances to the manner in which religious beliefs are held. The contents of the various beliefs differ: They may be about religious matters (creation) or metaphysical in nature (only observables count in science) or some scientific statement concerning the nature of reality (the existence of some tendency or force toward wholes and actualization present in reality or man), but the function they fulfil in science resembles the way in which religious convictions function in religious experience. Obviously, the term religious is now being used in a sense deviating from the standard and conventional usage. The fact that there are elements of this kind at work in theories has been dealt with quite extensively in the literature (Collingwood, 1940; Toulmin, 1972). This allegiance or commitment to a set of 'absolute presuppositions' or 'paradigmatic propositions' (Brown, H 1975; Brown, R. 1979) is a commitment held in very much the same way as fundamental religious beliefs. Perhaps this is what Lakatos (1974:139) would call the hard core of a research tradition. All problem statements, methodology and observations are developed in terms of these presuppositions. These statements called first order principles by Miller are used in the same way as religious statements, since they are statements in accordance with which all evidence is interpreted. Miller makes an even stronger claim:

> [...] basically religion and science present alternative Weltanschauungen, neither of whose first-order principles are amenable to empirical testing in that they actually function as principles in accordance with which all reasoning within the Weltanschauung takes place (1969:52).

What is regarded as 'facts' or evidence is determined by the *Weltanschauung* with its concepts, categories and first-order principles. Once this is understood, it is clear that science intrinsically harbours religious notions which are qualified by the characteristic context of science. Fiduciary components of science are not identical to fiduciary components of other domains of human life, but they do have an analogical resemblance to those realms.

Scientific concepts are embedded in a lexicon (Kuhn, 1989:9–32). Such a lexicon describes a possible world based on a network of beliefs. Metaphysical beliefs are as constitutive of the network of beliefs undergirding the 'possible world' as any other type of belief present in the network. These in turn are regulated, determined, and conditioned by a dominant control belief or set of control beliefs which determine the choice of ontological strata and conceptual schemata utilized in a theory. These ontological and conceptual schemata are closely related to (not causally determined by) a basic metaphor which dictates the choice of analogical emphasis present in the scientific metaphor functioning as control belief in the cluster of beliefs that make up a theory. The rejection of the Copernican heliocentric view of the world by the Roman Catholic Church is an example of an ecclesiastical religious belief ostensibly fulfilling an epistemic function within the weighing of a theory. I say 'ostensibly' because the actual regulating or controlling function within the Church's theory as such was fulfilled by the status and sanction awarded the geocentric Ptolemaic interpretation of the world. When the transition to the heliocentric view took place, it was ostensibly due to new 'scientific observations', but the controlling belief at work in the heliocentric view was the Neo-Platonic view of the centrality of the sun in the visible universe (Burtt, 1954).

Metaphors both reveal and conceal something of the nature of reality. In science the logic of the entailments of the metaphor prescribes the research program. In the set of entailments of the source domain of the metaphor the focus governs the mapping process. It is elevated to a position guiding/directing/controlling the search and what is 'seen' (perceived/observed). It sets up the proposed itinerary to trace the itinerary of meaning of the target domain – not the full target domain, only the itinerary indicated by the control belief of the source domain. A fact that also underscores the 'underdetermination thesis'.

Control Beliefs as Certitudinal Footholds of a Theory

Internal scientific control beliefs need not necessarily be religious in either origin or content in order to function religiously (i.e. analogous to religious belief) within the belief structure of scientific cognition. Any belief within the belief structure of scientific knowledge can be 'elevated' to the status of a control belief. The role of control beliefs is perhaps best illustrated by the different emphases found in Systems Theory (Bois, 1978:204ff; Boulding). These emphases correlate with the opening up of different analogical moments within reality. For example, the mechanical (Hobbes), organic (Durkheim), cybernetic elements of the system analogy gradually open up new perspectives on reality (Stafleu, 1980:26).

In each one of the mechanistic, organicistic and cybernetic versions of Systems theory, one qualifying dimension dictates the formation of the permissible lexicon. The clockwork image, growing plants and computers can all be regarded as systems. In turn, all three of these system images can function as a source domain to be mapped on a target domain. The meaning of the lexicon used in a theory is determined by the specific analogical element that constitutes the regulative focus of the metaphor 'system'. The radius of analogies of the metaphor inevitably carries with it its ontological and ideological freight.[2]

Scientific theories are embedded in a network of beliefs of which some are characteristically metaphysical beliefs (cf. Popper, 1965; Agassi, 1975; Botha, 1986a). Some of these beliefs are held in such a manner that one could characterize them as being the certitudinal foothold of a theory. Wherever scientific commitments are held in such an unquestionable sense or protected against possible falsification, they are candidates for the qualification as religious assumptions. Some might prefer to call them philosophical or metaphysical assumptions, but religious assumptions are actually philosophical assumptions of a certain kind. They are the type of religious assumptions that are held in such a way that they do not allow for their dependence on other assumptions. These types of assumptions do not only appear in the realm of organized religion in the form of doctrine or dogma, but analogous assumptions are also found in science, politics and most other areas of human life.

Metaphorical Hermeneutics a Methodology for the Natural Sciences?

It has become clear that so-called non-epistemic factors are often as epistemically constitutive of the process of scientific concept formation as the so-called epistemic factors (Nersessian, 1987; Tweney, 1989b). Moreover there is often a direct link between the basic metaphor chosen as a model of interpretation of reality and

2 Eubanks (1999) claims this more specifically for the notion of conceptual metaphor.

the type of root- and conveyance metaphors chosen in scientific theorizing. The metaphorical focus of a scientific theory becomes apparent when the control beliefs that are safeguarded by a theory are identified. The analysis of the structure of metaphor above provides a hermeneutic for both the social and the natural sciences. Moreover, it can also provide valuable insights into the relationship between religion, metaphysics and scientific theories. The analysis of Michael Faraday's use of the religious and scientific metaphors 'force' will demonstrate this thesis.

The function and presence of metaphor in natural scientific theorizing has been argued extensively (Bradie, 1980; Hoffmann, 1980; Hesse, 1966, 1972, 1974, 1975). In the last section of this chapter I would like to argue not only that metaphor is as constitutive of natural science as it is of social science and the humanities, a claim also defended extensively by others, but that natural science is as hermeneutic as the social sciences and humanities (Heelan, 1983a; 1983b and Hesse, 1980:185). And that comprehending the role of metaphor in natural science is a step toward the understanding of the constitutive role played by religious convictions in scientific theorizing.

According to Cantor, (1987:127): Boyd (1979), Hesse (1980), Harré and Martin, (1982) et al, metaphor performs three functions in science. The first is its *generative* function. This implies a far stronger role than only in the logic of discovery. Metaphors are theory-constitutive. Any worthwhile theory is constantly being extended through the articulation of its underlying metaphors. They are essential to the development of scientific theory. The second is the role it plays in the *framing of scientific terminology*. This has to do with the reference function of metaphor which functions both in immature and mature theories. These two functions thus operate at the level of theory generation and in research. The third role of metaphor has to do with *teaching*. Often metaphors are used in a didactic context in order to convey or demonstrate new insights in terms of already existing insights. Cantor (1987:128) shows that many of the optical metaphors used in eighteenth century light theory originated from theological discourse. The origin of such a metaphor does not make its function within the theory as such religious. What is of interest is the fact that scientific instrumentation and apparatus is often what Cantor (1987:140) calls 'a reification of metaphor'. The role metaphor plays in instrumentation and experiment is twofold: It is often exactly in so-called crucial experiments that the instrumentation is employed as an experimental extension of a metaphor in order to get at as yet unknown aspects of reality. Metaphorical extensions, Cantor argues, occur when the existing apparatus and instruments are used in non-standard situations. An example is the experimental attempts at 'weighing light' which was the implementation of the materialist metaphor for light.

A second more explicitly hermeneutic role is played when it is recognized that any experiment as reification of a theory constituted by a metaphor needs to be 'read' or 'interpreted' as any other text in either literature, science or reality in general within a horizon of meaning provided by the symbolic universe of discourse and by the parameters of the dominant disciplinary matrix. This is why Heelan can say: '[...] science is a hermeneutic of man-made signals produced in experimental situations and these get their meaning from and through the community of scientific enquirers' (1972c:495). This hermeneutical aspect is located in the heart of natural science: in observation and perception (Heelan, 1983a:181). But there is also a wider context which provides the context of meaning within which the instrumentation, apparatus and experiments as texts are read: This what Heidegger (1962) calls *Vorhabe*, i.e. the culturally acquired background of embodiments, skills and practices in which we hold/have the object of our understanding, even before we can recognize and name it as an object. It is metaphor which provides the connecting link between this *Vorhabe* of the wider culture and the context of meaning within which the scientist often works. For this reason it will also be one of the most fruitful avenues to provide access to the interrelationship between deeply imbedded cultural religious convictions and those scientific beliefs which are elevated to the status of religious control beliefs. Hesse writes:

> [...]theories have always been expressive of the myth or metaphysics of a society, and have therefore been part of the internal communication system of that society. Society interprets itself partly by means of its view of nature (1980:186).

It is the important constitutive role that metaphors play in such a view of reality or metaphysical framework and in theorising based on such a framework that opens new possibilities to interpret the relationship between religion and science.

Contrary to general opinion, religious factors are not per definition non-epistemic in nature. Their epistemic or non-epistemic character depends on the function they fulfil within the context of theorizing. Any scientific, philosophical or metaphysical notion can acquire a dual epistemic function in theorizing. It can function as the cognitive framework within which meaning is discerned and also as a control belief within a specific theory. The cited example of the transition from the geocentric to the heliocentric view of the universe is a case in point. It shows that a theoretical (perhaps philosophical?) notion can function as a 'religious' or fiduciary notion epistemically at work within the structure of the theory. Not only philosophical but also various other types of ideas and notions can, under certain circumstances, become epistemic in nature and even more specifically become epistemically-regulative and act as epistemic control beliefs. Acceptance of this approach will at least require revisiting the theory of meaning of scientific concepts.

Recapitulation

A view of the interrelationship between science and religion which does justice to the arguments advanced up to this point requires a theory of meaning which takes into account both epistemic and so-called non-epistemic factors in the process of concept formation. Such a theory of meaning avoids the pitfalls of two theses in traditional Philosophy of Science: viz. the radical distinction between theory and observation in science thesis and the radical theory-ladenness of observation thesis. Both theses operate with a reductionist theory of meaning in which the meaning of a theory is reduced to either one of the two poles. The proposal in this chapter has been to work towards a revised understanding of the nature of religion and the role of religious convictions in theorizing. A prerequisite for such a revised position is the acceptance of the fact that the two theses above are questionable. Moreover, metaphor and more specifically analogies play a constitutive role in theorizing and are intrinsically bound to religious factors of both the epistemic and the non-epistemic kind. In the following chapter, the arguments developed here will be illustrated by the interrelationship between religion, metaphysics and science in the work of Michael Faraday.

VIII Religion, Metaphysics and Science in Michael Faraday's Work

The cognitive historical approach to the history of science provides a methodological approach which makes it possible to take into account most of the factors which are constitutive in the formation of scientific meaning in scientific concept formation. Here this approach is coupled with a methodology of metaphorical hermeneutics in order to demonstrate the role played by religious factors in the meaning of scientific concepts. This method is applied to a case study of the relationship of faith and science in the work of the nineteenth century British physicist, Michael Faraday. It will be used to analyze the relationship between religion, metaphysics and science in Michael Faraday's work, analyze the change in conceptual structure which the 'force' concept undergoes in his work and demonstrate the role of fiduciary control beliefs in his work.

Most writers on Faraday's views about matter agree that his religion was a significant factor in his work. The fact that these religious convictions operate via a set of metaphysical principles is part of the Cantor thesis (1991). Views differ when it comes to the question whether his religious convictions only shaped the broad parameters of his thinking or were effective in shaping the content of his theories, so do opinions as to the extent of the influence of Kantian idealism and *Naturphilosophie* in Faraday's metaphysics. Both Faraday and Newton shared a common allegiance to the Judeo-Christian religion, yet their metaphysical beliefs and their final choice of theories for the understanding of matter diverge. An interesting question raised by this is: Why do ostensibly similar religious beliefs foster diverse metaphysical and regulative principles in scientific work? I will attempt to provide a provisional answer to this question.

Faith and Science: Methodological Approach

In any attempt to give an account of the relationship between science and religion the philosopher and historian of science is confronted with the important methodological issue of clarifying the central terms used in his research. The basic definitions of religion and science will determine how the relationship is construed and explained. Religious influence is often defined as the presence of religious or theological themata in scientific work or when religious and theological ideas are seen as the source for inspiration of scientific theories. Scholars often derive their ideas from religion or theology or are influenced by theological or religious notions in the process of theory formation. A much-cited example of such an approach from the history of science is that of James Clerk Maxwell, the nineteenth century physicist in whose scholarly work the notion of the Trinity is claimed to have played a pivotal role (Torrance, 1984:215–242). Such a methodological approach thus argues that if a scientist were guided by ideas which have their origin in Scripture, faith or theology, this would be proof of the influence of faith in theorizing. Here faith, the Bible or theology are seen and used as the source of the theory formation. Although the history of science abounds with examples of this kind, careful analysis soon shows that such sources, origins, or themata do not necessarily make the resultant theories Christian or religious.[1] Moreover, it is often the case that such so-called religious notions are based on suspect theology and exegesis at times influenced by non-Christian philosophical influences.

In the body of publications in History and Philosophy of Science, on the other hand, there is growing evidence of the recognition of the variety of possible ways that religious and anti-religious convictions actually do influence scientific work. According to Rudwick:

> [...] these beliefs are no longer regarded as embarrassing for the reputation of a great scientist, nor are they dismissed as irrelevant to the 'real' work of science in past periods. Furthermore, they are no longer treated only as 'external' factors that in particular circumstances may have retarded or advanced the intrinsic 'internal' development of valid scientific conclusions. A growing number of historians of science now routinely treat religious beliefs as a possibly constitutive factor in the formulation and acceptance not only of generalized world-views, but also of quite specific scientific theories and even of alleged observational 'facts'. Religious beliefs, in other words, are taken seriously in the task of understanding not only the context but also the content of the scientific work of past periods (1981:244,5).

In the Philosophy of Science literature there have always been defenders of the thesis that philosophical and metaphysical elements influence theorizing in some or other way (cf. Burtt, 1952:17; Agassi, 1975:208; Botha, 1986a:430, etc.), but this influence is often restricted to the context of discovery or the heuristic phase of theorizing. A stronger thesis, to be defended here, claims that not only do philosophical and metaphysical assumptions and notions influence science constitutively, but that these metaphysical and philosophical assumptions are often the expression of reli-

1 Cf. the important works of Funkestein (1986) and Brooke (1991).

gious influence which penetrates scientific work via (religious) control beliefs on the level of experimentation, empirical work and concept formation. The idea that metaphysical and philosophical notions penetrate scientific work to this level is often regarded as a contentious claim. The reasons for this are the ambiguous meanings of the term religion and the persistent legacy of the positivist prejudice against the recognition of the role or presence of metaphysics in science. I plan to defend a variant of the stronger thesis, one in which I will argue that philosophical and/or metaphysical frameworks harbour control beliefs that fulfil a regulative function in the development of scientific theories and the generation of meaning of the key terms of a theory or research tradition.

The contentiousness of the stronger thesis is also dependent upon differences of opinion concerning the formation of the meaning and reference of scientific concepts. Positions found in the literature differ: The positivist distinction between theory and observation language would ascribe primacy to literal language and deny the presence of metaphysical notions in theorizing. On the other hand, there is the position of the proponents of radical meaning variance such as Feyerabend et al who argue that all observation is theory laden. I have already argued that both these positions are questionable. An intermediate position is one in which the role of metaphorical language and models in theorizing is acknowledged. Such an intermediate position is that of Nersessian (1987) who defends a network theory of meaning. She argues that both extreme approaches to meaning and reference of scientific concepts are reductionistic and inadequate to explain how the generation of meaning takes place in scientific concept formation. Nersessian (1987:164) points to the fact that accounts of meaning and meaning change in scientific theories have left out actual scientific practices and the human beings that invent them To rectify this she makes a methodological proposal to fully incorporate the dimension of discovery in the philosophical analysis (Nersessian, 1992:6,41). She calls this approach a cognitive historical approach to meaning which investigates those processes through which vague speculations are articulated in new theories and gradually come to replace the existing scientific representations in a specific domain. The theory of meaning she proposes (1984a; 1984b) must take into consideration all those factors influential in the generation and formation of meaning in theoretical concepts. This entails that factors regarded as non-epistemic based on older theories of meaning, are now regarded as epistemic and actually contributing to the process of scientific meaning formation.[2]

An intermediate position concerning the meaning of scientific concepts is also that of Stafleu (1980:27) who argues that in the dilemma of meaning variance and meaning relativism one needs to defend a position which does justice to both. He argues that in the historical process of the opening up of a field of science the meaning of the concepts used are gradually more clearly defined, leading to what Kuhn would call the acceptance of a paradigm and mature science. But as scientific revolutions develop the stability of meaning of these concepts are challenged. The meaning of the concepts change, yet retain some constant core meaning which makes communication possible. Ultimately the concepts are embedded in a metaphysical framework which is the result of the 'ideology' or worldview at work in the paradigm.

An argument in defence of the stronger thesis will be developed here, incorporating some insights from Nersessian's (1987) proposed cognitive-historical approach and the Philosophy of the Cosmonomic Idea of Herman Dooyeweerd. I argue that world views operate via meta-theoretical frameworks in scientific theoriz-

2 The important discussion between Nersessian and Leplin (1991b:678-686) in this regard will not be dealt with here, although it certainly could provide important insights for the problem under discussion.

ing. They influence scientific concept formation and meaning in empirical and experimental work via the regulative function of control beliefs embodied in such frameworks. Contrary to views that define religion on the basis of the object of belief or belief in divinity or transcendence, I argue that in some metaphysical frameworks certain absolute presuppositions are awarded such priority that their function in a scientific theory is analogous to the manner in which a religious belief functions for a believer (Miller, 1969:64ff). Because the basic metaphor foundational to a metaphysical framework mediates abstraction, it is often an indicator of the type of control beliefs awarded priority status in the metaphors and analogies generated by the basic metaphor. It is exactly these control beliefs that reveal the philosophical presuppositions which function in a religious manner in theorizing. It is to the role of these beliefs and presuppositions in the work of Michael Faraday that we now turn.

Faraday and the Elusive 'Forces' of Nature

The publication of Cantor's (1991) work on Faraday has contributed to the general discussion of the relationship of faith and science. Cantor (1991:167ff) argues that Faraday's science was permeated with religious (although not explicitly biblical) meaning. He argues that Faraday's theology of nature and his understanding of the economy of nature comes to expression in a scientific vocabulary in which he attempts to articulate his religious conviction that the physical world was created according to God's plan. Cantor argues that these terms constitute 'metascientific principles' which sometimes functioned as necessary truths and sometimes as guidelines in his scientific theorizing and ultimately in his research (Cantor, 1991:169).

The fascinating search which brought Faraday to the development of a philosophy of nature and matter that diverged fundamentally from his predecessors spans a lifetime of experimental work in Chemistry, Physics, Electricity, Magnetism and Electromagnetism. Central to all these discoveries was Faraday's force concept – a concept he imbued with new meaning. Most biographers of Faraday agree that his religious convictions played a decisive role in these discoveries and specifically in the meaning he ascribed to the concept of 'force'. This meaning contributed to a radically new understanding of forces and fields in the natural sciences. Faraday asks in his (1816) unpublished, so-called Commonplace Book[3]:

> 'Questions': 'Bodies do not act where they are not' Query – is not the reverse true? Do not all bodies act where they are not; and do not any of them act where they are?' (Tweney, 1989a:97).

The problem that matter could not act where it is actually not present was one of the questions which already received attention before the seventeenth century (Hesse, 1955:337). From Faraday's earliest notebooks, it is possible to conclude that he too was immensely preoccupied with the problem of the understanding of the forces of nature. He was intrigued by questions such as:

3 Faraday, M. 1816. Commonplace Book. Unpublished Manuscript in the Collection of the Institute of Electrical Engineers, London: England.

- How the earth could attract the moon when there was clearly no mediating agent between these two entities?
- How does a magnet attract a nail?
- Why do some charges attract whilst others repel?

These questions posed problems simply because in ordinary experience attraction and repelling was experienced in the sense of push and pull. This ruled out the possibility that matter could act where it is not.

The dominant worldview of Faraday's times was still the Newtonian with its understanding of action-at-a-distance. Because Newtonianism harboured elements of both the mechanistic and the dynamic interpretation of matter, Newton is often appropriated in the history of science by both traditions, although it is clear, according to Levere (1971:60) that it is the dynamic aspects which dominate Newton's view of matter. Faraday distanced himself from the mechanistic emphases of the Newtonian tradition.

In his view of matter, Faraday distanced himself from three important views of the Newtonian tradition: the notion that there are hard, impenetrable particles, the existence of empty space between these particles and the existence of centrally directed forces which would act upon forces at a distance. Many scientists had difficulty with Newton's idea of action-at-a-distance, but Faraday rejected this notion and replaced it with an understanding of the nature of matter which was far more dynamic (I would like to say 'dynamistic') in nature. This he does through the introduction of a central concept force. The seventeenth century idea that matter consisted of atoms of particles was already questioned in the eighteenth century by various scientists: Both the idea of contact-action of particles and the idea of action-at-a-distance was rejected and in its place variations on the theme of the existence of some 'universal stuff' with varying degrees of intensity, mostly called 'ether', were developed. When Faraday develops his force concept there were three possible models available in terms of which matter could be understood (Hesse, 1955:338):

- Contact-action (impact).
- Action-at-a-distance.
- Action mediated through some or other medium.

The way matter was seen at the time was still much closer to the mechanistic understanding. Faraday changed this state of affairs with his ability to use physical and pictorial representations to represent the behaviour of matter. Opinions differ as to whether this change in Faraday's understanding of matter dates from a later developmental phase of his work, after 1844, or whether it was latent in his earliest views. Heimann (1971) talks about a 'conceptual dichotomy' in Faraday's earlier conceptions about '[...] particulate polarization, and his later work which was founded on the concept of the primacy of the lines of force' (1971:237ff). According to Heimann (1971) two opposing views are present in Faraday's work: a particulate theory of matter and molecular polarization on the one hand and on the other his later lines-of-force notion. Heimann (1971) shares Hesse's (1961:199) point of view that Faraday's *1844 A Speculation Touching Electric Conduction and the Nature of Matter* represents a transition from a more mechanistic to a dynamistic view. She writes:

> The transition was foreshadowed by Boscovich and Kant, but Boscovich does not equate his matter with its force, and still regards his theory in terms of action at a distance, while Kant, through equating matter with repulsive force distinguishes between this, which acts by contact, and attractive force which acts at a distance (1961:201).

Nersessian (1984b) on the other hand argues that the idea that electrical or magnetic action at a distance does not take place directly, but requires time and therefore implies the existence of space between the molecules or conducting medium, was already present in Faraday's work as early as 1832. She thus believes that the field concept or the alternative force concept was already latent in his much earlier work. It was through experimenting with electrical induction that Faraday discovered that the action between the molecules did not take place at a distance, but through some or other medium. This medium was not a material medium, but something else. His experimental work in this regard led him to believe that some or other active medium existed which conducted electricity and magnetism. This active medium he represents with the lines-of-force which is the embryo of the field concept later refined by Maxwell and Einstein.

According to Cantor (1991) Faraday's religious convictions played a decisive role in his quest to understand the nature of matter. In general I agree with this position but even though this is the case, one can discern two different philosophical positions in his work with respect to the notion of force. It is the relationship of his religious convictions to these two differing views on force to that are of interest here.

Faraday's Field Concept

The representation which Faraday made of this medium was the well-known lines-of-force or magnetic curves (Anderson, 2006), of the electro-magnetic field (the pattern formed when iron filings were strewn in a magnetic field). The lines of force were not represented as material entities, but as lines which behaved like bundles of elastic: they could expand, bend, go around curves, manifest tension, collapse and vibrate. The line character of this force provides the analogy which forms the basis for the description and explanation of the behaviour of matter and it is specifically the movement of lines which are significant. Faraday sees static and dynamic electricity as the expansion and breakdown of the lines; magnetism, light and gravitation as the vibration of the lines and matter where the lines converge (Nersessian, 1987:170). The forces of nature all form a unity for Faraday because they reflect the unity of the invisible Power of God (Williams, 1965; Berkson, 1974; Gooding, 1980, 1982; Cantor, 1985, 1991). These forces are not properties of matter, but are manifestations of this one unified force.[4]

The alternative views of matter in his time were materialism (atoms as impenetrable particles) and point atomism which saw atoms as centres of force (Boscovich, 1966). Both Kant and Boscovich saw matter as such points where forces converge. Many authors argue that Faraday was influenced by Kant and Boscovich (Spencer, 1967/68; Berkson, 1974; Jammer, 1957; Williams, 1965:78) but claim that he deviates from their points of view to come to the conclusion that matter as such does not exist but that there are only forces. His *A Speculation Touching Electrical Cconduction and the Nature of Matter* (1844a), *Matter* (1844b) *and On the Physical Lines of Magnetic Force* (1852) are most probably the best illustration of this point of view. Both Heimann (1971) and Hesse (1961) see these publications of Faraday as representative of a fundamental change in his view of force. Faraday argues that the properties which matter possesses can only be known and distinguished based on the forces operating in matter. For example gravitation

4 This view is of course more representative of his views after 1848. Cf. his second force schema.

or weight is recognizable on the basis of the forces of attraction; hardness is known on the basis of the repulsion which an object exhibits; all that we know about matter and its operations, can only be known on the basis of the effects of forces operating in matter and these forces operate in every single atom of the molecule up to the minutest point. His conclusion is: Why could God not just as well create forces around points or centres. Why would He have to create a nucleus of matter and then endow it with power or force? In a later argument, Faraday (Faraday, 1844:853,4; Levere, 1968) goes one step further and takes leave of the notion that there could be some hard point or centre around which the forces are grouped. This leads him to the conclusion that force is the only physical substance. His force concept eradicated the difference between matter and space, because it combined the notion of forces active near matter and in matter itself into one concept. A development I would like to argue is representative of a move towards a full blown dynamistic understanding of matter. From a position that force is one of the manifestations of matter, he moves to a position where matter and force are identified. What role, if any, did his religious convictions, worldview or metaphysical framework play in this transition?

The Relationship of Faith and Science in Michael Faraday's Physics

Differing positions and interpretations are found in the literature concerning the relationship of faith and science in Faraday's work. These differences pertain to his philosophy of nature, the developmental phases that can be distinguished in his ideas and the possible influence that his religious convictions might have had on these ideas. Although there are differences of opinion on the precise relationship between his religious convictions, metaphysical framework, meta-theoretical principles and his experimental work, there seems to be a convergence of thought concerning Faraday's understanding of nature. Two force schemata are at work: a Newtonian-mechanistic one which Heimann (1971) calls a 'particulate theory of matter' and a more dynamic, even dynamistic understanding of force (Nersessian, 1984a; Tweney, 1989b:356; Heimann, 1971:238–9; Hesse, 1961:201). The juxtaposition of the mechanistic and dynamistic views and the qualification of the former as Newtonian is misleading to some extent, because both traditions in the philosophy of matter can be traced back to Newton. Here the adjective Newtonian will primarily be associated with a particulate and atomistic view of matter. The dynamistic understanding of nature expounded by Faraday is a stronger one than the dynamic views of matter found in the Newtonian approach.

As already noted there are differences of opinion regarding the question whether these two force schemata are already present in Faraday's earliest work and whether the dynamistic view is representative of a later development. Views also diverge on the possible influence that his Sandemanian religious convictions could have had on his views of matter in general and the transition in his views of force in particular. With respect to the views concerning faith and science, there are those who claim that Faraday held to a strong separation between faith and science, a position for which some support can be found in his writings. Others, like Cantor (1991;1996) argue that his religious convictions fundamentally influenced the metaphysical framework of his work. In his work Faraday's literal reading of Scripture and his empiricist epistemology, with its requirement of the exclusion of all prejudice and speculation in the literal reading of nature, played a pivotal role. Between these two positions there is a point of view such as that of Heimann (1971:245) who argues that although Faraday's religious convictions contain general claims about God's power and the order of creation, it does not necessarily entail any specific theory about or view of nature and matter. Cantor's work on Faraday sheds much light on the symbiosis of religion and science in Faraday's work.

I agree that the presence of religious convictions in Faraday's scientific work become apparent in his metaphysical and metascientific principles, but I believe that their influence should more specifically be sought in the influence of religious control beliefs at work in his theories. These become apparent in the semantic foci of the basic analogies operating in Faraday's two force schemata. For the purposes of this analysis, I will take my lead from Cantor's excellent exposition of the relationship between Faraday's religious and scientific convictions. However, I do have some reservations. My definition of 'religion' differs quite fundamentally from the one used by Cantor. I do not believe that he would share the thesis that there are two metaphysical schemes at work in Faraday's understanding of 'force'. The consequences that I claim follow from this thesis and Faraday's literal reading of the power of God are obviously for my own account.

Faraday's Religious Convictions: The Cantor Thesis

In Cantor's chapter (1996) *Interpreting Michael Faraday as the Christian Philosopher: The Problem of Metaphysics* he shows that a symbiotic relationship existed between Faraday's religious, metaphysical and scientific ideas. The religious convictions developed in his theology of nature influenced his scientific work through the mediation of a set of meta-theoretical principles. Faraday believed in the economy of nature – a metaphysical conviction which was in itself inadequate to articulate his physical theories. According to Cantor (1991) his metaphysical assumptions included the following: Nature is God's creation, Scripture is literally (1991:193) true; the invisible power of God sustains all the forces of reality; all forces of nature are the expression of this invisible power of God, etc.

From these metaphysical principles, Cantor argues, Faraday proceeded to draw several cashable '[...] metascientific principles [...]' including principles such as: The unity and simplicity of nature, the power of God must be understood in a physical sense, physical power is maintained and conserved and is indestructible, the lawlikeness of nature and the conservation and correlation of powers, the principle of plenitude and the economic distribution of the lines of force (Cantor, 1991:168–174). Cantor also shows that there are several instances where the relationship between religious convictions and scientific insights are demonstrated in Faraday's work. Possibly the following reflection from Faraday is one of the most pertinent:

> And when we [...] perceive that all this is done by virtue of powers in the molecules which are indestructible, and by ways of action the most simple and unchangeable, we may well, if I may say it without irreverence, join awe and trembling with joy and gladness. Our philosophy, feeble as it is, gives us to see in every particle of matter, a centre of force reaching to an infinite distance: binding worlds and suns together, and unchangeable in its permanency. Around this same particle we see grouped the powers of the various phenomena of nature: [...] the harmonious working of all these forces in nature, until at last the molecule rises up in accordance with the mighty purpose ordained for it, and plays its part in the gift of life itself. And therefore our philosophy, whilst it shows us these things, should lead us to think of Him who hath wrought them; for it is said by an authority far above even that which these works present, that the invisible things of

Him from the creation of the world are clearly seen, being understood by the things that are made, even His eternal power and Godhead. (Levere, 1971:76,7).

From this it is clear that for Faraday there existed an intimate correlation between the powers of nature and the invisible power of God (Cantor, 1996:6). It is exactly this religious notion of power in Faraday's understanding of Romans 1:20 that had a decisive influence on the way he developed his concepts of power (and force). Let us return to the Cantor thesis in which he shows that two factors played a decisive role in Faraday's work: the literal hermeneutics of Scripture prescribed by his religious affiliation to the Sandemanians and his empiricistic epistemology which implied that just as Scripture ought to be read without prejudice or hypothesis, so too God's revelation in creation ought to be understood as literally as possible (ibid.,1991:196ff; 295).

Faraday was a member of the Sandemanians, a religious sect that broke away from the Church of Scotland because it differed fundamentally about the Biblical justification for the existence of a national church. Their religious convictions were strict and fundamentalist. They did not interpret Romans 1:20 in the tradition of natural theology as if it would be possible for man to come to the knowledge of God via the knowledge of nature, but, on the contrary, they argued that it was the knowledge of God which made it possible for man to understand nature. Levere writes:

> [...] he adopted a theology of nature, rather than natural theology. This inability of man to find God by 'reasoning from the things of this life' also comes to the fore at the commencement of his lecture 'On mental education' in 1854 (Faraday, M. 1859. (1991). *Experimental Researches in Chemistry and Physics*, 464) Faraday says: 'Let no one suppose for a moment that the self-education I am about to commend in respect of the things of this life, extends to any considerations of the hope set before us, *as if man by reasoning could find out God*' (Italics mine) (1971:78.9).

The reference to Romans 1:20, which is found repeatedly in Faraday's work, served as a central religious thread of the metaphysical framework that permeated his work. Faraday could not interpret the notion of the 'power of God' metaphorically, because he believed that God did not reveal Himself in a metaphorical fashion but spoke clearly and literally. Power was for Faraday '[...] the very stuff of the physical world' (Cantor, 1991:175). Every single particle of reality was a manifestation of God's power and this power came to expression in heat, electricity, magnetism, gravitation, chemical affinity, etc. But for Faraday electricity was the highest of God's powers of the universe. For Faraday the Sandemanian, the threads of God's power run through all aspects of human life and through the physical world, created by God. He rejects the image developed by natural theology in which God is seen as a skilled craftsman or artist. Instead of forming the world like an artist, God created reality through His word. Faraday believed that it is not possible to know the origin of these powers of the universe and that it is also not possible to know how theses powers are united – this is known only by God. All that man can know is the effects of this power and the relationships between these forces and the laws that regulate the effects. According to Cantor (1991:185) Faraday's argument was that God created a perfect system at creation in which no loss of power as the result of physical processes could take place. The total amount of force in reality is maintained, cannot be destroyed and cannot be recreated. Humankind has no access to the absolute power ('power force') in the universe. All that we can know are the observable forces of nature like heat and electricity. It is only God that can create or destroy power. The abstract entity 'power' exists behind the individual observable manifestations of power, but man has no access to this power. Faraday wrote:

> [...] we know nothing of [the] origin of powers or how they are united with matter, we are outside the legitimate bounds of science if we try to specify power or force as an abstract entity. Since all powers in the physical universe are manifested by God speculations on the nature of force may be encroaching into the area of theologically dangerous hypotheses. Put simply, while correlation concerned the relation between observables, conservation was a metaphysically charged term. (Cantor, 1991:187).[5]

Cantor (1991) argues that both Joule and Faraday held to a theology of nature in which it would be absurd to think that the powers that God created in matter could in any way be destroyed or created through the creative activity of man.

> [...] such a belief in the 'economy of nature' was subject to a number of different and even incompatible interpretations. For example, Joule and Thomson who subscribed to latitudinarian principles and to natural theology, envisaged their God conserving the mechanical operations of nature in accordance with the principle of conservation of work and 'vis viva' (living force). Joule could speak of 'the wisdom and beneficence of the Great Architect of nature' and of 'the entire machinery' of the universe working 'smoothly and harmoniously' – terms from natural theology that connote God as the Designer of a mechanical universe. By contrast as a Sandemanian, Faraday interpreted the Bible in a more literal manner and portrayed the God of Genesis as the source of power, which is conserved. The powers we observe in matter are ultimately derived from that power but they are not reducible to mathematics or mechanics. Faraday's God was no artificer or engineer (1991:192–3).

Faraday thus talks about the Divine Economy of Nature which comes to expression in the indestructibility and conservation of power, but not of energy. It is exactly his notion and understanding of force and power which prevented him from accepting the

5 Perhaps this is one of the strongest indications as to nature of a metaphysical conviction which functions as a religious control belief in his work.

later mechanical concepts of power, work and energy developed by the energy physics of Joule, Thomson et al. Faraday's understanding of the conservation of power is not that of the first law of thermodynamics. Joule e.g. argued that heat could be transformed into power and vice versa, but Faraday saw this in direct opposition to his deepest religious conviction that motion was not one of the properties of matter. Such a view would imply that an ordinary physical phenomenon could create power and this privilege was God's alone.

It is obviously not possible to do justice to the exhaustive analyses of Michael Faraday's work found in Cantor's tome. The Cantor thesis is a valuable elucidation of the relationship between Faraday's religious and confessional convictions and their influence via the mediation of metaphysical principles and meta-theoretical assumptions. From the analysis one can gather that Faraday's religious beliefs allowed two metaphysical frames and two force schemata in his work, one closer to the classic Newtonian mechanistic understanding of force and the other already harbouring elements of a more dynamic understanding of force which only came to fruition in the work of later physicists. The Cantor thesis does not shed much light on the possible influence of Faraday's confessional religious beliefs, specifically the interpretation of Romans 1:20, on the transition from a more mechanistic to a more dynamistic force conception. Cantor does identify the metaphysical and meta scientific principles operating in Faraday's work, but does not fully address the constitutive role of religious-like beliefs in scientific theorizing. So, we are still faced with the question as to how one set of confessional religious beliefs would allow for two metaphysical frameworks and two force schemata.

One Set of Confessional Religious Beliefs, Two Metaphysical Frameworks and Two Force Schemata?

Various possibilities present themselves here: It is possible that it was exactly his Sandemanian beliefs with their strong emphasis on the dynamic power of God, that directed his work toward a more dynamistic understanding of matter. But this would imply that these beliefs mainly influenced the latter developments of his work. Such an approach does not take into account that his religious beliefs were present and operative from the very beginning of his work. They were already adhered to when he initially embarked on his career and developed his work in the light of a more mechanistic and particulate understanding of matter and nature. One could also resolve the problem by arguing like Nersessian (1984) that the dynamistic view which borders on the later-to-be-developed field theory actually was latent in his work from the very beginning and that there is not adequate evidence to speak of a transition here. So the question remains whether his religious convictions changed through the course of his career or whether one set of religious convictions actually sanctioned two force schemata operating in his work.

There are indications of a development from the more mechanistic to a dynamistic view, which implies that one set of confessional, religious convictions (the 'power of God' in Romans 1:20) actually allowed and or sanctioned two sets of metaphysical assumptions (the mechanistic and the dynamistic). The Cantor thesis gives an explanation for the fact that Faraday interpreted the force concept literally, but does not explain the role of religious convictions in the transition from the one force schema to the other. What actually seems to have been the case is that one set of confessional, religious convictions (inter alia the literal reading of Romans 1:20) sanctioned the literal interpretation of two different force schemata. This presents some problems to a proponent of the strong thesis concerning religious beliefs and scientific theorizing that argues for a positive relationship between religious convictions, metaphysical principles and scientific work. Can one still argue that religious factors are at work in his scientific work, when it is clear that they obviously allow for divergent metaphysical views and theoretical notions? Cantor's analysis of Faraday's work poses a problem for any argument in favour of a direct and possibly causal relationship between confessional, religious factors and scientific work. The Cantor thesis or any general thesis that religious convictions influence scientific theorizing presents us with the challenge to show how this is constitutively the case.

One possible strategy to reach a solution to the problem posed above is to restrict the proposed analysis to the relationship of metaphysics and theorizing and to abstract from the role of religious convictions in the conventional sense of the word. Such an approach is one in which for example the possible influence of Kant or German *Naturphilosophie* is held responsible for the dynamistic tendencies in Faraday's metaphysics. I would contend that even when abstracting from the possible influence of conventionally conceived religious convictions and concentrating on the relationship of metaphysics to theorizing, religious factors[6] would still be at work. They would be 'religious' in the sense of being the philosophical or metaphysical expressions of absolute presuppositions about the nature of matter This would then be 'religion' in the revised sense of the word.

Many problems remain to be dealt with. To note a few: With regard to the possible influence of the Sandemanian injunction to

6 'Religion' is here defined in the sense of an ultimate commitment to that which is non-dependant. In science it reveals itself as absolute presuppositions functioning in a regulative or controlling fashion.

read the Bible literally the question remains: What was prior? Did Faraday's literal reading of nature with its strong emphasis on the literal understanding of force give rise to a literal physically dynamistic understanding of God's power in Scripture or was the influence the other way round? What does a literal reading actually mean? Obviously, it presupposes the distinction between literal and non-literal or metaphorical. How would a literal mechanistic understanding of force differ from a metaphorical reading of force? How do literal and metaphorical readings of a mechanistic and dynamistic interpretation of force differ? Is it possible that force could be understood both mechanically and dynamically and in both cases, the specifically qualified notion of force could be understood either literally or metaphorically? These are the sorts of conundrums that can be addressed by the principle of aspectual universality which was explained in an earlier chapter.

The Force Metaphor: Two Literal Readings?

The transition to and introduction of Faraday's dynamistic conception of matter implements a new representation of force. If one argues that the particulate understanding of matter represents a literal understanding of force, then the introduction of the new notion of force represents a new metaphorical representation based on a pictorial representation in which the lines-of-force analogy plays a basic role. In this analogy, a different analogical element of the force metaphor is given priority status. In his early understanding of force Faraday worked with the analogy: Matter is, acts as if it is, (like) fields of force, i.e. force is seen as one of the properties of matter. In his experiments in which a current- carrying wire was made to rotate in a magnetic field and even in the earth's magnetic field, Faraday writes: 'It appears the action does not regard the material substance at the centre but simply the current of electricity or the magnetic pole, the substance merely giving locality to the power' (Faraday's Diary 1932–36, vol. 1:93 Entry on January 21, 1823). Tweney (1989a:98) thinks that this comment implies an important advance in his thought: Material substances provide a locality for the operation of forces, but forces have an independent reality and are not reducible to mere attributes of matter.

In his quest to understand the nature of this irreducible force and under the influence of his metaphysics, Faraday later dissolves matter into force and works with the identity of matter and lines of force. The non-literal analogy or metaphor is transposed into a literal similarity: Matter is (like) moving physical lines (of force). This notion of the lines of force is laden with a meaning which in principle excludes the possibility of it being material in nature, yet his metaphysics prescribes both reality and particularity to the notion. The attributes of the image (of the lines of force) are projected onto the notion of a field, according to Tweney.

> Faraday's methodology and his field concepts are in fact different manifestations of a single developing world view, a product of Faraday's fundamental reliance on a procedural epistemology fully consistent with his ontological assumptions. For Faraday, knowledge was anchored in concrete action; 'abstract' knowledge was a secondary, derived entity (1989a: 92).

In the same way, physical reality for Faraday was anchored in concrete forces. Matter was a secondary, derived entity. It is this connotation of the field and lines of force which are informed via his religiously based metaphysics and his scientific understanding of power and force.

Faraday's field conception involves the notion that electric and magnetic actions actually take place along or through curved lines of force. The curvature of the lines was an essential component of his argument that these actions are not actions-at-a-distance; he attempted many times to demonstrate the curvature experimentally. Faraday's efforts focussed on trying to formulate a coherent conception of how it was that the motion of the lines, could account for the forces of nature. Nersessian demonstrates the way the idea is implemented in Faraday by describing the different activities of the lines of force: they would expand and collapse, electrostatic induction might take place through the action of contiguous particles along curved lines of force, whereas light and gravitation could be the result of a 'shaking' or 'vibrating' of the lines. Even matter itself might be nothing more than point centres of converging lines of force (Nersessian, 1987:170).

The image of the lines of force is a conveyance metaphor that provided Faraday with terminology with which to discuss and reason about the phenomena and to communicate his interpretation to others. In discussing the actions, he used such terms as 'moving out', 'expanding', 'collapsing', 'bending', straining', vibrating', 'being cut' and 'turning' corners. The concrete analogies he chose to describe the type of action he had in mind were 'line-like' phenomena such as rays of light and heat, rings in water, and conduction through wire. The line-like phenomena acted like 'curves' which he interpreted as the 'motion' of the lines. Because this movement manifested itself 'in space' outside (and inside) matter, Faraday concluded that there is only force; that matter is nothing but force. The analogies postulated in two force schemata are the following:

Force Schema I: Atomism:

Analogy 1: Matter is like (electronic) contiguous particles

Conceptual Structure of Schema I: Atomism

Matter is an entity.

Matter (is like) (particulate forces)

Example of a force: Electromagnetic induction is due to a tension of contiguous particles of matter: The electrotonic state of the polarized state of the intervening particles.

Force is a property of the entity matter.

Extension, gravitation, electricity and magnetism are instances or manifestations of this property.

There is no action-at-a-distance, but particles convey force (up to ½ inch [...])[7]

7 Cf. Heimann, 1971:239ff; Faraday, 18 April 1840. Cf also Williams, 1965: Selected correspondence of Michael Faraday, vol. 1, 1812-1848, 360ff; Tweney [1989b:356]).

Force Schema II: Dynamism

Analogy II: Force (is like) moving curved lines in space and in matter.

Conceptual Structure of Schema II

Matter is abstract.

The powers are concrete.

There can be no abstract notion of force.

Matter is force.

Force is an entity.

Faraday thus focuses on one property of matter, viz. force, represents it with a spatial image endowed with the characteristic of movement; observes that this property is present in and outside of matter and concludes that this property is the all pervasive physical reality. Force, which is a property of the physical phenomena he experimented with, is now seen as identical with matter and is not regarded as a property of matter but as an entity. This gives rise to the identification of matter and force and the obliteration of the distinction between an entity (matter) and (one of) its properties, force. Faraday reified the curved lines of force.

> For Faraday, the curved patterns of force became real curves, immaterial but substantial entities that can be cut by a moving conductor (thus inducing a current), can be considered as moving or at rest, and can be considered as possessing numerosity – the more curves cut, the greater the induced current (Tweney1989a:99).

What happens in the second force schema is the reification of force: Matter IS moving, curved lines of force in space. The intriguing question is why this change takes place? The Cantor thesis argues that Faraday's confessional religious convictions were instrumental in his views and by implication also in this change. A strong argument could be developed that the change in analogical focus was also brought about by the influence of the German idealistic *Naturphilosophie* to which Davey was inclined. Faraday, Davey's assistant, did not read German. The Royal Institution in England, where he worked most of his life, possessed very few German works. It is possible that he could have been exposed to the ideas of German Romanticism via the influence of Davey who was an avid admirer of the English poet Coleridge (Williams, 1965:69;1973:17). There is also the probable influence of Kantian notions. Portions of *The Philosophy of Nature* can almost literally be found in Faraday's work (Levere, 1981; Kant, 1786).

A more balanced reading of Faraday would take both the confessionally religious influence and the influence of the *zeitgeist* into account. A cognitive-historical approach would call for such a comprehensive approach, but the emphasis here is on the change in religious control beliefs represented by the change of focus in the underlying analogy chosen by Faraday to explain the forces at work in matter. The literal reading of nature through the mechanistic and particulate concept of force is replaced by a dynamic understanding of matter via a new metaphorical focus of force. The injunction to read force literally in this case leads to an over-extension of the force metaphor. This is brought about by the change in control beliefs operative in his work.

Let us return for a moment to Faraday's later argument concerning the nature of matter. Reconstructed, the argument reads as follows: There is nothing but force in reality. He concentrates on one of the aspects of matter, viz. power, visualizes it by means of a spatial image in which the force is represented as moving lines of force, observes that this phenomenon is not only restricted to matter, but is also present in the space between particles of matter and then concludes that there must exist such an all pervasive force, the lines-of-force. In the metaphor 'matter is (like) moving lines of force or fields of force', matter is equated with force. The element of dynamic movement of lines acquires the dominant focus in the analogy, to such an extent, that Faraday concludes that all that exists is force!

Other cultural ideas and practices might also have been influential in this literalising of the metaphor. Cantor has pointed to the influence of the requirement of Plain Style and Plain Language in the hermeneutics of Scripture at the time[8]. Moreover, the prevalent double language thesis with its emphasis on the literal might also have contributed to Faraday becoming a victim of the metaphor of force. Possibly, it could be concluded that Faraday was victimized by the metaphor of force so that his force concept is in actual fact a form of metaphorical hypertrophy (Weltman, 1973)[9] brought about on the one hand by his deep religious convictions, but also mediated by the scientific control beliefs central to his metaphysics. What Jammer says about the later rule of the concept of force in the history of physics, could perhaps very aptly be applied to

8 Cf. Glas, 1811.

9 Stafleu, 1987:249 speaks of theoretical hypostatization

Faraday's understanding of force: '[...] the concept of force rose almost to the status of an almighty potentate of totalitarian rule over the phenomena [...]' (1957:241).

This critical rendering of Faraday's use of Scripture in the process of experimental work poses the challenge to indicate what scholarship would look like when the metaphors of Scripture do function responsibly in theorizing. What is of utmost importance is to show how Scriptural metaphors function hermeneutically in our understanding of reality and how they relate to the hermeneutical choice of metaphors to access or explain reality. Aspects of these questions will be dealt with in the next chapter.

IX Clusters, Roots and Hierarchies of Metaphors in Scripture and the Quest for Christian Scholarship

Scriptural, Disciplinary and Reality Metaphors

Recognizing the multi-dimensionality of reality and the plurivocity of meaning and signification on which we rely in both literal and metaphorical language use assumes the existence of non-linguistic and pre-conceptual bases which guide the recognition of similarities, differences and analogies in reality. They in turn are pointers to a God-given creational design plan for reality. Metaphorical meaning appeals to and presupposes such an ordered and categorized world to which language and texts refer. These categorizations and classes provide limits and boundaries to the multiplicity of deferrals of meaning that inter-textual relationships seem to imply.

That there is such an order is primarily a matter of faith, but it also becomes apparent in the constraints that reality imposes on conceptual meaning. Deeply embedded experiential conceptual metaphors (Lakoff, 1993:202–3); Kövecses, (2002; 2005) is an indication of how closely our language habits are related to the way we imagine the world. Moreover, metaphor provides access to the discipline's assumptions about how the world is structured. Elsewhere I have argued that theory-constitutive metaphorical language harbours and mediates access to the structure of the world (Botha, 1993a:60–73). This chapter is an elaboration of this thesis and an attempt to differentiate the diverse types of Biblical metaphors and indicate how they relate to theory-constitutive metaphors in some disciplines.

Here I explore how the confessional thrust of Scriptural metaphors and root-metaphors relate to the scholar's quest and how choices for hermeneutical keys to Scripture relate to metaphorical keys chosen as access to reality. With respect to Biblical interpretation, it is argued that the text of Scripture itself provides the theologian or reader with leads concerning the type of metaphorical access that functions as its hermeneutical key. I argue that there are clusters or hierarchies of metaphors, central and root-metaphors, which regulate the interpretation of Scriptural texts and that a redemptive historical reading of Scripture as a confessional text guides the meaning of such clusters and hierarchies of metaphors. The root-metaphors in Scripture set the certitudinal parameters for the metaphors chosen and utilized in the disciplines.

In the disciplines, metaphorical models play a hermeneutical role in the understanding and interpretation of reality. In these metaphorical models, control beliefs steer, guide and condition the access of the discipline to reality. For the scholar who takes the role of Scripture seriously, the control beliefs embedded in ultimate presuppositions, need to comport with the thrust of Scripture and its root-metaphors. Scholars are at work in God's creation and their metaphorical approximations of the structures of this creation are guided by the contours of created reality. These approximations in turn are influenced and constrained by that to which they attribute ultimacy in the process of attempting to understand this reality. The stable God-given order provides the conditions and parameters for the common differentiation of contexts within which the interpretation of God's Word in creation is to take place.

The central question is how the scholar should attempt to relate her choices for Biblical points of departure and her Biblical world view to the pursuit of theoretical knowledge concerning the structures for and in God's world and what the relationship is between Biblical metaphors and the metaphors utilised in theories in the disciplines.

Biblical Metaphors as the Source of Theorizing?

My hypothesis is that the relationship between Scripture and science is mediated by two different sets of metaphors – Scriptural metaphors and the metaphors utilised in the theories of the disciplines. Together they approximate the analogical stratification of reality. Historically there have been diverse ways in which scholars have attempted to relate Scriptural insights and images to scholarship (Brooke, 1991:16–51). The possibilities listed by Brooke do not only hold for religious convictions, but also for the role of Biblical notions in scholarship.

Some Christian scholars derive their ideas from religion or theology and are often influenced by theological or religious notions in the process of theory formation. This approach argues that if a scientist is guided by or influenced by ideas which have their origin in Scripture, faith or theology, then this would be proof of the influence of faith in theorizing (Torrance, 1984:215–242). Here faith, the Bible, or theology is seen and used as the source of the theory formation. Although the history of science abounds with examples of this kind, careful analysis soon shows that such sources do not necessarily make the theories Christian or religious. To the extent that theorizing takes the Biblical narrative about creation, fall and redemption seriously, it approximates scholarship which could be qualified as authentically Christian scholarship. Such scholarship, rooted in God's covenantal love for His world and His gift and call to all His creatures to obey Him, will produce perspectives that shed light on the path of scholarship and are conducive to growth of insight into the nature of culture, history and human relationships. Due to its confessional focus and nature, the Biblical narrative does not prescribe or proscribe the content and substance of theories with which to approach or explain dimensions of reality. It does not provide the metaphors that can function as explanatory models in scholars' theories about the world. It does, however, set their direction and focus to the religious relationship to that which is regarded as ultimate, God. In this guiding and direction-setting function, there are different kinds of Scriptural metaphors, root-metaphors and clusters of metaphors at work. In these metaphors, a certain hierarchy can be discerned. Biblical metaphors need to be articulated in a wider metaphysical framework or Biblical worldview in order to be of use in the theoretical work of scholars in the disciplines. This approach needs to be differentiated from those that urge a return to Scripture as the basic precondition in the limited sense below.

Back to Scripture?

In the work of a number of recent Christian scholars, one finds a call to return to the Scriptures as a prerequisite for the development of the Christian scholarship enterprise. This is a call with which I can resonate in many ways. Wolterstorff's[1] proposals to relate Scripture and learning have elaborated central themes of the Gospel, *justice and peace* and situated them in the midst of contemporary societal issues (Wolterstorff, 1983 & 1984). Others propose to resurrect the notion of *care* (Goudzwaard & De Lange, 1991:72–73) and an ethos of compassion (Hart, 1995:67–96) in social relations, to mention but a few. In the epistemological stalemates posed by naive realism and radical constructivism some have proposed the consideration of an epistemology of *stewardship* which emphasizes gift and call, i.e., a relational epistemology '[...] committed to respecting the other, attending to how the other discloses itself to us'[…][2] (Walsh and Middleton, 1992:167–171) or to replace the idea of knowledge as power and knowledge as control with the understanding of knowledge as *intimacy* (Jennings, 1997:124) for us to come to know and love others. Knowledge with love according to Jennings will mean the transformation of the current symmetries of production, reproduction, and arrangement of knowledge.

These worthy proposals have one refrain in common, namely a return to and application of the very concrete claims of Scripture to the way society is structured and also to the way we form knowledge and theories of social reality. These proposals are also characterized by the fact that they do not propose the search for Scriptural principles or the recognition of creational ordinances as

1 Wolterstorff (1983; 1984) deals with both epistemological and social articulations of this central issue. In both books the notion of "justice-in-shalom" is central to the articulation of the relationship.

2 The South African philosopher H.G. Stoker introduces the notion of "fanerosis". He advocates an epistemology which recognises the intrinsic revelational (fanerotic) character of reality.

the way to go[3], neither do they argue the need for a metaphysical framework to mediate these Biblical notions in scholarship.

But, these approaches raise some questions: Why choose the one Scriptural dimension and not the other? Why emphasize love, or compassion, or community, or intimacy, or care, or justice, or peace? Do they not all have to come into the full-orbed image of human lives in obedience to God? Why privilege the one Biblical emphasis over the other? Moreover, do not all these Biblical emphases also require philosophical articulation in order to become fruitful in the enterprise of the academy? One needs to recognize the multi-valency of God's law both in Scripture and in creation and to honour the multi-dimensionality of its validity. There is an element of truth in postmodern insights into the pluriformity of the nature of the world we inhabit. There is also an element of truth concerning the wide spectrum of possible ways of coming to grips with the nature of the world.

All signification has as prerequisite the existence of a stable and reliable structural order which makes it possible and conditions the broader process of knowing the ordered and patterned nature of the world and God's law. This structural order is that to which the metaphorical models in science and metaphorical language refer. It is this structure which is articulated by the theoretician and conditions the intentions of the author or actor and the bias of the interpreter. Both Scripture and the disciplines provide a view of reality and both these approaches utilize metaphors.

3 See Griffioen (1997). See also the analysis of this historical development in Botha (2000a).

Metaphor as Views of the World in Scripture and in the Disciplines

Not only are metaphors constitutive of the world views made by groups and cultures, they are also constitutive of the world views construed by academics and theorists (Kuhn, 1973; 1974:459–517 and 1979).[4] They form part of the elements with which human language habits imagine and construct a world view. To what extent such metaphorically based theoretical world views in the disciplines are informed by, depend upon or are conditioned by Biblical insights and views of the world, is a critical issue for scholars who are serious about Biblically informed scholarship in the disciplines. They are as dependent on answers to questions concerning the role, place and meaning of metaphor in Biblical and religious language as are their counterparts in Biblical scholarship. It is the relationship between root-metaphors found in Scripture (e.g., creation, fall, redemption, covenant, kingdom, law, etc.), those chosen as hermeneutical keys to interpret Scripture (e.g., narrative, liberation, feminist, structuralist, Marxist, historical-critical approaches, etc.) and those developed in the disciplines (e.g., mechanisms, organisms, systems, conflict, drama, play, etc.) to gain access to the nature, structure and functioning of some dimension of reality, that poses the challenge to the Christian scholar. How the metaphors of everyday language, of religious experience, and of theoretical models relate to the metaphors in Scripture is the focal question. This requires exploring how the confessional thrust of Scriptural metaphors and root-metaphors condition and constrain the parameters of theoretical meaning within which terms and concepts function in the Christian scholar's quest. Implicit in

4 Cf. Postman's use of World makers (1996:172, 177).

this exploration is the question how choices for hermeneutical keys to Scripture relate to hermeneutical keys chosen as access to reality.

Disciplinary and theoretical approaches to subject matter are often embedded in perspectival overviews of reality. This is apparent in the pervasive world views that permeate the perspectives of theories (Clouser, 1991:66). In these approaches a root-metaphor often provides the theoretical vocabulary (permissible metaphors and analogies) based on which explanatory hypotheses are developed and pursued. Such root-metaphors can consist of a selection of one or more aspects of reality which are assumed to provide a framework in terms of which some aspect or all of reality can be understood. It is the relationship between root-metaphors found in Scripture and those chosen as hermeneutical keys to interpret Scripture, and those in the disciplines which are used to gain access to the nature, structure, and functioning of some dimension of reality, that is a central issue for Christian scholars. Most taxonomies of philosophical frameworks of theories in the Social Sciences do their analyses based on the identification of key notions [Cf. Altman and Rogoff (1987:7–40), Brown (1977) and Stein (1963)].

Not only do philosophical choices concerning the nature of language influence how Christian scholars will in actual fact integrate faith and learning, but it will also condition the way scholars approach their subject matter. This is especially true for the way they give an account of the ultimate grounding of the categories, classifications, differences and similarities which they regard basic to the nature of the world and which their language attempts to articulate. These are philosophical issues that entail positions on realism (Botha, 1986a), reference and truth, both in an ontological and epistemological sense.

The Omnipresence of Metaphor in Scripture

The metaphorical nature of the language of the Bible is indisputable. In this respect it does not differ from the nature of ordinary language or from the language used in science and theology. Scripture is replete with well-known metaphors about God, His revelation in Scripture and in nature and about believers. God is portrayed as 'Creator', 'Father', 'King', 'Judge', 'Shepherd', 'Redeemer', etc. Scripture is depicted as the 'Word of God' and a 'light' by which the light is actually seen (Psalm 36). Christians are described as the 'body of Christ', a 'temple', the 'children' of God, 'friends' of God, 'slaves', 'ambassadors', 'light', 'salt', 'fishers', 'letters', etc. These and many other entity-like metaphors and metaphorical utterances are to be found in Scripture. Scripture also utilises other kinds of anthropomorphic metaphorical images which relate to aspects or facets of reality and God's existence: God is love, He is justice, He is righteousness, etc. Both types of metaphorical utterances are woven into the warp and woof of God's revelation to humankind. Scripture as a book of revelation utilizes both kinds of anthropomorphic images to portray and represent God, human experiences of their relationship to God and experiences of His world. An important question is whether one can discern a hierarchy in these Scriptural metaphors.

With respect to Biblical interpretation the text of Scripture itself provides the theologian or reader with leads concerning the type of metaphorical access that functions as its hermeneutical key. There are clusters or hierarchies of metaphors, central and root-metaphors, which regulate the interpretation of Scriptural texts. One such dominant cluster is the feminine image of wisdom in the Old Testament wisdom literature (cf. Perdue, 1994 and Gammie & Perdue, 1990). A far more encompassing root-metaphor on the other hand is the notion of redemption of the fallen creation. A reading of Scripture as a confessional text in the light of this central root-metaphorical notion guides the meaning of all other clusters and hierarchies of omnipresent metaphors in Scripture.

Often it is argued that religious utterances are dependent on metaphor because of the difficulties related to establishing the 'reality' of the referents in these fields. It is argued that metaphorical language indicates the inability of the believer to fully approximate, grasp and articulate the reality of the transcendent God who escapes full human comprehension (Van Huyssteen, 1997). No doubt, this is true, but a similar argument is also possible with respect to fully grasping or articulating much else in reality. Therefore, for example, the reality of justice, love, communication, etc., always transcends and escapes a full grasp by human language. Ubiquitous metaphorical language is indicative of this fact. In this sense, the way metaphorical language is utilized in and about Scripture does not differ fundamentally from ordinary and everyday language even though the transcendent God as referent obviously does. Yet, it is common knowledge that religious utterances which are replete with metaphorical language have a unique character, which distinguishes them from other language use and more specifically other uses of metaphorical language in poetry, myths, or therapy. It is the unique and typical character of this religious type of metaphorical language and its relevance for metaphorically mediated world views and scholarship that requires closer scrutiny. A few examples should suffice at this stage.

Scripture often speaks about the intimate relationship between the (Biblical) notions of justice and shalom. Obviously, these terms are used in contextually qualified situations in Scripture, but the central-religious thrust of all Biblical terms such as justice and shalom are comprehensive in scope. Epistemologists, legal scholars and specialists in conflict resolution too are deeply interested in notions of peace and justice. Their sense of justice and peace could be closely related to the comprehensive Biblical notions, but not synonymous with these Biblical notions. Their use of justice and peace relates to a specifically qualified form of justice and peace. In a similar fashion the epistemological claim that Christian theories are those that pursue justice-in-shalom as Wolterstorff (1984) claims is related to the full and comprehensive Biblical meaning of the terms, but requires unpacking what *theoretical justice-in-shalom* is all about. It requires an articulation of the consequences of these Biblical or certitudinal metaphors for theory. Legal scholars who appropriate biblical notions of justice as bases for their legal theories can not escape the task of indicating and articulating how justice in the legal sense differs from political and economic justice and how these two forms of differentiated justice in turn relate to the full, and total Biblical notion of justice. When appeals to justice-in-shalom are made with respect to societal issues it requires the theoretical articulation of a philosophy of society which can comport with a Biblical understanding of society and the way in which peace and justice can be articulated in a 'societal' sense. This means that shalom and justice in the full religious and Biblical confessional sense of the Word is more than its differentiated articulations in diverse concrete situations.

In their recent book *The Gift of the Stranger: Faith, Hospitality and Foreign Language Learning*, Smith and Carvill (2000) claim that there are also limited metaphors that are relative to the specific pursuits of scholars. They do not reject the role of over-arching narratives or root-metaphors but unpack the limited metaphor of hospitality to the stranger. They argue that Scripture gives a significant position to the stranger in our midst and requires hospitality to be part of the life of the Christian community. This Biblical image or metaphor should form the basis of the teaching of foreign language, they argue. This type of approach should be welcomed as a more sophisticated hermeneutical approach to the use of Scripture than fundamentalist or Biblicist proof texting but it raises some interesting questions about the use of Scripture in Christian scholarship. Is this the way that religious beliefs or convictions ought to enter the intrinsic structure of theorizing? How does one relate Biblical teachings and metaphors to the nature of 'language' as such? Does the choice for a specific metaphor as a basis for a specialized discipline not divorce the limited Biblical injunctions chosen as the basis of theory from the overall thrust of the Biblical message? To illustrate the point: Could an appeal to the Biblical injunction of hospitality to the stranger not also justify other theoretical positions such as sanctioning an open immigration policy at the expense of the citizens of a country, or a refugee policy of some kind?

The point is that a Biblical metaphor often allows a wide diversity of possible theoretical positions, but is characterized by the fact that it rules out, prevents or prohibits certain positions or assumptions. Clusters of metaphors or individual metaphors still need to be read against the background of the central redemptive historical thrust of Scripture. The Biblical injunction or metaphor is a pointer, a compass, not a road map. It sets direction, delimits parameters but does not dictate the exact way to go. Moreover the direction it dictates acquires its calibration from the central thrust of Scripture – a thrust which can be expressed in a central or root-metaphor. Theoretical work requires far more detailed and elucidated theoretical approaches and articulation of world views and philosophical frameworks than can be provided by or gleaned from these (theoretically) limited metaphorical Biblical insights. Theoretical scholarship requires an approach to scholarship conditioned by what can be regarded as the root-metaphors of Scripture. The critical question is which Biblical metaphors are decisive and central to his project.

Metaphorical Keys to Scripture and Reality

As stated earlier, metaphorically constituted theory language harbours and mediates the underlying root-metaphorical assumptions about the nature of the 'possible worlds' (Kuhn, 1989:9–32) imagined by theories. This is true too, of theories that attempt to articulate views of reality within the parameters of the confessional thrust and content of the Biblical message (cf. Sinnema, 1975; Olthuis, 1987). On the level of meta-discourse[5] about the discursive practices in Scripture the question arises whether it is possible to discern one or more central, traditional, privileged, or classic (root) metaphors that regulate both the pre-understanding of Scripture in the life of the church and the theologies that are imaginatively generated to give account of this pre-understanding. Van Leeuwen speaks of the 'nucleus symbol' (1990:112) of specific sections or books in Scripture that would by nature of their status capture that which is essentially the intent and purpose of God's revelation concerning Himself, His plan of redemption, or the nature of His reign. Such (root) metaphors, being central themes of Scripture, would dictate the manner in which Scripture ought to be read or understood. They express those insights that have been regarded

5 Metaphorical language is often employed as a meta-discourse to describe and interpret discursive practices found in Scripture (to quote Gaonkar [1997] who makes this distinction in a different context). Thus, for example, a marxist, feminist or structuralist meta-discourse can be employed in an attempt to give an adequate interpretation of the discursive practices in Scripture. In such meta-discourses, a choice is often made for some metaphor (often originating in Scripture, but often also imported into Scripture) in terms of which the whole corpus of Scripture is read, interpreted and understood. This metaphor functions as a hermeneutical key or type of logic-of-discovery or the exploration of Scripture.

as permanent and essential components (McGrath, 1998:185) of the truths of the Christian tradition. This could be called the root-metaphor question.

The relevance and significance of choices for or against the primacy of specific Biblical images, events, symbols, or metaphors as hermeneutical keys to the understanding of Scripture are also decisive in the relationship between Scripture and scholarship. Choosing the centrality of the redemption in Christ, (Van Huyssteen, 1997:144) or covenant and kingdom or law leads to different interpretations, as would a choice for the primacy of the notion of creation (Stoker, 1967; Wolters, 1985) or the Word (Olthuis, 1976:88). Others (Spykman, 1992; Bartholomew, 1995:11, 32) differ on the choice of a specific central or primary theme in Scripture. Bartholomew says if one sees the Bible as an edifice with many entrances, then covenant is one of the main entrances that give us a unified and overall picture (1995:32). Similar insights which take the notion of God's covenantal law as a central notion and also the key to the understanding of creation formed the basis of the Dutch philosopher Herman Dooyeweerd's Christian philosophy which was developed on the basis of seminal insights formulated by the Dutch theologian Abraham Kuyper[6]. The original title of the Dutch version of his philosophy was De *Wijsbegeerte der Wetsidee* (The Philosophy of the Cosmonomic or Law Idea) set the notion of law central to this philosophy.[7]

Various attempts have been made to characterize the essential message of the text of Scripture as a whole.[8] 'The Story-line of the

6 Dooyeweerd would not agree that these are metaphorical notions. Cf. Chapter IV of this work and also the discussion in Botha (2000a).

7 More recently Hart (1995:67-96) and others have criticized Dooyeweerd's one-sided choice for the notion of law and have argued that it has led to a legalistic and logicistic interpretation of Scripture.

8 Perhaps one of the most recent examples of such an interaction is the

Bible' found in the Introduction of the Bible Society version of the Bible *Into the Light* and Wolters' (1985) *Creation Regained* comes to mind. Obviously, these choices beg the question of whether the nature of Scripture is such that one can actually choose these metaphors or whether one has no choice but to accept those metaphors that are given (McGrath, 1998:185).

I agree with those who make a strong case for the fact that creation must be understood in terms of the covenantal law which governs its existence. This covenantal law is the merciful and trustworthy relationship between God and His Kingdom creation. These encompassing notions recognize the fallen nature of this creation and the redemption and restoration of creation. Whether we choose a central metaphor as point of entry to understand the 'holistic grammar' of the whole text of Scripture or whether we come to discern the governing metaphors of specific texts within Scripture or just identify an inventory of common metaphors to be found in Scripture, Scriptural metaphors have one thing in common: they are certitudinally or confessionally qualified. This is what Olthius calls '[...] the overriding focus of Scripture' (1987: 86).

It is possible to develop solid arguments for a large number of possible points of entry to God's revelation in Scripture, as it is possible to access creation from a diversity of points of entry. The choice for any one of these Scriptural metaphors and root-metaphors as hermeneutical keys will have a different impact on the Christian scholar's quest and access to reality. There is a correlation between most central Biblical notions like justice, love, righteousness, care, frugality, stewardship, discernment and compas-

---

choice of theologians during the era of apartheid in South Africa to interpret the Biblical message from the vantage point of God-given creational ordinances with the emphasis on ethnic and cultural separation and diversity.

sion in the full Biblical sense of the word and the concrete expression of love as it presents itself in a diversity of aspects of human experience. These Biblical themes are not enclosed in a single restricted meaning, but have a multivocity of meanings which are qualified by their depth-dimensional religious meaning. These Biblical metaphors play a guiding, controlling and directing role in human life and the understanding of reality and do not dictate the contents of theorizing.

Central or Privileged Metaphors?

Which Biblical metaphors are candidates for the most likely and most biblically responsible root construal of the central reality of Christianity? Those which have traditionally been regarded as the classic or traditional Biblical metaphors have been challenged by McFague (1985 and 1987) and others. McFague argues that the basic metaphors and images used to interpret faith for our times have remained relatively constant: triumphalist, monarchical and patriarchal. These she claims, are outmoded and oppressive metaphors and models, and ought to be replaced by a remythologizing of the relationship between God and the world. She then experiments with the models of God as mother, lover and friend of the world and with the image of the world as God's body (McFague, 1987:65, 66, 71). Here the source of the metaphors is not Scripture, but contemporary human experiences. Her analysis is provocative and worthy of consideration, albeit only because it is serious about the injustices perpetrated on women and the world in the name of a patriarchal reading of Scripture. Whether her representation of the patriarchal, triumphalist understanding of God is fair is open to question. Even more questionable are her alternative panentheistic metaphors that are read into Scripture by means of a feminist hermeneutics and then exegeted out of Scripture using a similar methodology – a dubious move to say the least. Perhaps the single most questionable move in her feminist hermeneutics is replacing the Biblical understanding of God as King and His Kingdom with a pantheistic or panentheistic reading.

Yet much needs to be said about the pervasive and overarching meta-narrative of Scripture which illuminates all human experiences and provides perspective on the origin, nature and ultimate goal of this world. An integral historical-redemptive reading of Scripture in which reality is portrayed as created by God, fallen and broken through human hands and redeemed in and through Jesus Christ, provides the indispensable understanding of both Scripture and reality. Here the central metaphor is the creative Word of God, at work in creation, coming to expression in Scripture and becoming incarnate in Jesus Christ.

Religious Language Between the Poles of Pan-Literalism and Pan-Metaphoricism

Positions concerning the nature of religious language and more specifically the language of Scripture can be mapped out on a continuum between two extremes: Pan-metaphoricism on the one hand and pan-literalism on the other. Pan-literalism would simply state that although the Bible makes use of metaphorical language these metaphors could ultimately be reduced to literal language. Not all authors would agree to a pan-metaphoricist reading of everyday language or the language of Theology and Science. Some would claim this privileged position of metaphor only for the language of faith, religion and Scripture. Therefore, for example, Polkinghorne (1991:2) claims that metaphor and symbol are par excellence the language of Scripture, theology, and religion, whereas mathematics is the natural language of physics and science. Such a view is often combined with one that argues that metaphors are our only access to Scripture or that all talk of God is metaphorical in nature (McFague, 1987:34). This type of pan-metaphoricist approach needs to be distinguished from arguments such as those of Hesse (1983:40) and others who argue that '[...] all language is metaphorical' – a position sometimes also dubbed pan-metaphoricist. The former position rests on the double language thesis and then claims that metaphor is our only access to spiritual and religious reality. Hesse's (1983) claim is a revision of the double language thesis. She recognises that within specific contexts one is still required to work with this distinction, but then argues that the distinction is not only contextually determined but also relative and embedded in meanings that are not cast in natural kind categories but fluid and changing, variant and dynamic. Her network theory of meaning acknowledges the cognitive claims of metaphor.[9]

The arguments of Van Huyssteen (1997) and McFague (1987:33) about God talk being irreducibly metaphorical are the

9 It is difficult to identify the central figure who initiated the notion that the double-language thesis was questionable. Lakoff (1993:202-203) attributes it to the seminal essay of Reddy (1993) 'The Conduit Metaphor'. In Philosophy of Science, this issue surfaced within the context of the theory-ladeness of observation discussions in the wake of the Popper-Kuhn developments. The transition from the classical double-language thesis to the far stronger position that 'all language is metaphorical' had also already been introduced in the very early publications of Hesse and was an elaboration of the earlier reflections of Richards and Black on the 'interactive' nature of metaphor.

type of argument that could be called theological pan-metaphoricism, one of the many species of pan-metaphoricism. This theological pan-metaphoricist claim attributes primacy to metaphorical language. Van Huyssteen (1997:87) sees literal language being replaced by an awareness of the metaphoricity and relationality of all language, but within the framework of postmodernism it is especially religious language, he claims, that is characterized by this change. He writes that religious narrative leads us to see through the window of metaphor, to the way we ought to believe. Epistemic access through metaphor is therefore neither empirically deductive, nor literally true, nor subjective illusion. Such a position is disputed by Van Woudenberg who argues that it is untenable mainly because '[...] in principle it is possible that what is said with metaphor can be transposed into literal speech' (1998:231–247) – a statement that could be construed as pan-literalism. Perhaps the most critical question raised in the case of a choice for a pan-metaphoricist position is what the implications of this position are for the interpretation of Scripture. Does a pan-metaphoricist understanding of the nature of Biblical language necessarily commit one to the rejection of the historicity of the events being described in Scripture or to scepticism with respect to the reality of God (Wright, 2006:191)? Does it of necessity imply textual indeterminacy? In this respect Soskice notes '[...] this debate centres not on whether religious language is ineradicably metaphorical, but what follows if this is so' (1987:108; cf. 1985). When language and specifically religious language is seen as metaphorical, it implies that this language has as much cognitive import as so-called literal language. The nature of its cognitive weight is qualified by the fact that it is a unique confessional or certitudinal way of speaking about the world, human life and its relationship to God. The question that follows from this is whether Biblical metaphors are 'literally' true.

In an earlier chapter I argued that the double language thesis which forms the backdrop to the question just asked, be replaced by an understanding of the contextual qualification of what is regarded as literal and metaphorical. I hesitate to use the language in which the literality of Biblical truths is questioned for the reason just stated.[10]

The 'Literal' Truth and Biblical Metaphor

Biblical and religious language, which is intrinsically certitudinal and confessional, relates to the concrete, experiential relationships of the community of faith and expresses their way of seeing reality. In these utterances or statements, metaphors mediate the construction of a religious perspective on the world. Everyday religious metaphors such as the statement *The Lord is My Shepherd* are part and parcel of the confessional and certitudinal expression of a way of looking at and being in the world. They are time- and history-bound and as such are very closely related to the system of categories and classifications characteristic of the specific culture. And yet, exactly because a metaphorical expression is utilized to express this deep religious and certitudinal insight, it remains valid and true in vastly different settings.

When such religious and Biblical metaphors function within highly theoretical and abstract contexts e.g. in theological reflection, it fulfills a similar constructive role. It selects, organizes and focuses the attention on a specific dimension of Scripture. The

10 I refer to the example of the "literal" meaning of the 23rd Psalm which is a confessional meaning.

choice for the root-metaphor of liberation in Liberation Theology is a case in point. Between the everyday uses of religious metaphors, the contextualized qualification of the meaning of Biblical metaphors, and their theological utilization there is both commonality and difference. The metaphor of God's Guiding Hand expresses some basic human metaphorical insight into and belief concerning the nature of God's guidance and Providence in human life. The same notion of the Hand of God can be abstracted from the overall thrust of the broad scope of the Scriptural narrative in order to utilize it as the pivotal theoretical key for the development of a theological doctrine concerning God's providence. Then it functions in a different, highly theoretical fashion, often as explanatory model. These issues are compounded when scholars claim to do scholarship in the light of the Word of God. They are confronted with the additional question of how the chosen theological models for the interpretation of Scripture relate to the metaphors and models chosen to approximate the structure of phenomena in reality. The outcome of the debate on issues of this kind is crucial to the understanding of the authority of Scripture within the church, theology, and the academy.

The multivocity of meaning of texts and plurality of perspectives mediated by metaphor is at the root of many debates dealing with plurality of perspectives and indeterminacy of textual meaning (Poythress, 1987) in postmodernism. Both theological realists and theological instrumentalists recognize the presence and role of metaphorical language yet differ fundamentally concerning the question whether this metaphorical language actually refers to a cosmos transcending divinity in the traditional sense (Soskice, 1987:108). While many of the issues raised in the debates about determinacy or indeterminacy of texts remain on the level of language, Soskice introduces the reality depiction of metaphor into the discussion. Reflection on metaphor in religious texts and religious language inevitably invokes the intersection of both sets of issues, namely the semantic and ontological which in turn raises the issue of limits and conditions of meaning. Hermeneutics in ordinary interpretation of Scripture and its theoretical theological counterparts are both bound to the overall itinerary of meaning of the overarching Biblical story In this itinerary of meaning Biblical root-metaphors fulfill a direction setting function.

The wide-ranging debates on metaphor in most other disciplines have clearly shown that the ostensibly linguistic issue concerning reference and meaning of metaphor is more than just that. It is inextricably embedded in epistemological and ontological issues, like matters of truth, reference, and meaning. A simple choice for the primacy of the metaphorical or the primacy of the literal has consequences for Biblical interpretation, translation, and theological reflection. The outcomes of such choices in turn resonate in other areas of life. The most salient example of this is probably the literal reading of the creation narrative in Genesis with its inevitable impact on Christian scholars committed to the implementation of such a literal reading in their natural scientific explanations of the origins of the earth. In both cases a literal reading of the Biblical account or the claim of a scientific theory to being literally and factually true, ignores the fact that both forms of interpretation lean heavily on the mechanism of metaphor to convey its meaning and that the respective metaphorical meanings are in turn conditioned by contextual factors. It is the contextual qualification and function of language which determine metaphoricity and/or literality.

There is no primordial uncontaminated literal language that can form the basis or the final court of appeal when it comes to meaning ascription. As Hesse notes: 'All language, including ordinary descriptive language, is metaphorical in the sense that its use of general terms implies a normative qualification of the vastly various multiplicity of things' (1983:40). This means that both what has traditionally been called 'literal' and 'metaphorical' language depends on similarities and differences that are present in the categorizations of reality. The approximations of the categorizations change over time. Moreover what is regarded as literal and metaphorical also changes in the course of history.

Some Concluding Reflections

Metaphors make and remake the world by classifying and categorizing it in certain ways. What at some point in time is regarded as the literal language in a specific vocabulary is as much tied to a system of categories or family resemblances as is metaphorical language (Hesse, 1985/86:32). At the root of both literal and metaphorical language is the basic (ontological) structure of differences and similarities on which all predication rests. This being the case, literal language deserves as much attention as metaphorical language has acquired[11]. What at a given time is regarded as the 'plain sense' or the literal sense of a text is bound to the function it fulfils in a community (Tanner, 1987). So-called 'literal descriptions' too, are stratified and categorized (Hesse, 1983:32). Literal language is as dependent on classification and categorization as metaphorical language and both are dependent on some prior (I would like to argue ontological) stratification[12]. This being the case it is clear that

11 Lakoff (1986:291-296) and Leddy (1995:205-222) have attempted to disentangle the multiplicity of meanings of literal language.

12 At this point I refrain from addressing the important issue of the relationship between essences, so-called natural kinds and the ontological structure of reality. It will be dealt with elsewhere.

an appeal to literal language in Biblical exegesis is inadequate to stem the tide of relativism ushered in by postmodernism. The solution needs to be sought elsewhere. Such a solution would at least recognize the relativity and contextuality of the literal-metaphorical description, the decisive nature of the contexts within which metaphorical utterances function and the cognitive claims of metaphorical language. It would also need to accommodate the fact that many concepts harbour a potential multiplicity of meanings.

Biblical scholars often appeal to the literal sense or plain sense of Scripture in response to postmodern attacks that question essentialist and foundationalist claims to knowledge. Such a position often presupposes a reductionist notion of metaphor in which metaphor is viewed primarily as an aesthetic or decorative device and is seen as ultimately reducible to literal meaning. Opting for the literal sense of texts or for the primacy of the literal, avoids addressing the issue of how both literal and metaphorical sense is both context-bound and conditioned by conventional categorization and social consensus. The reason for this is simple: literality is as problematic an issue as metaphoricity, not only for Bible interpreters but also for interpreters of reality. What is at stake in relating Scripture to the work in the disciplines is discerning what the typical character of the Biblical narrative is all about and how it prescribes the parameters and limits of translation and interpretation of both literal and metaphorical Biblical language. God's inscripturated Word[13] is addressed to reality but in a unique, certitudinal fashion. It also entails the recognition that Scripture as religious and confessional text speaks to the whole or totality of human life, whereas the disciplines usually have a far more limited and theoretically restricted focus. The disciplines too have their

13 Olthuis (1976) distinguishes between the Creational Word of God, the Inscripturated Word in Scripture and the Word Incarnate, Jesus Christ.

own typically specialized narratives coloured by the contexts which they investigate. The challenge for the Christian scholar who professes that God's revelation is normative for her own work is that some clear and meaningful relationship between metaphor-mediated Scriptural truth and the metaphor- mediated truth about the world needs to be established. This obviously entails a theory of metaphor which does justice to what is conventionally called literal meaning both in Scripture and in statements about the nature of reality. It also entails a clear understanding of the fundamental difference and relationship between Scriptural (confessional/certitudinal) language and theoretical, disciplinary language. It is especially the prophetic critique which Biblically conditioned metaphorical language can bring about in its confrontation with the contemporary mythological metaphorical rhetoric which will show its decisive significance.

Part Four

Metaphor and Its Moorings

X Metaphorical Models and Scientific Realism

Metaphor in science faces similar challenges to those faced by the so-called 'double language thesis' in everyday language use. In science, the correlate of this double language thesis is the legacy of the positivist distinction between observational and theoretical language. Theses which attempt to reduce metaphorical to literal meaning are the equivalent in everyday language to the scientists' attempt to reduce theoretical to observational language. In both cases Black's suggestion of the 'interaction view' of metaphor brought about a thorough rethinking. Not only was it clear that the distinction in both areas was porous, fluid and contextually determined, but also that the role of metaphor in science was far more constitutive to science than had been envisaged under the reign of positivism. Aspects of this problem were addressed in the chapter that dealt with the role of the fiduciary element in scientific theorizing. A few more interrelated issues need to be addressed. The first concerns the cognitive claims of metaphor and the second the relationship between metaphorical models and scientific realism.

The Cognitive Claims of Metaphor

That metaphorical re-descriptions of the primary domain of the explanandum in science leads to a deepening of understanding and an extension of insight into certain structural aspects, must be attributed to the 'new way of seeing' made possible by the metaphor.

This not only constitutes a figurative or symbolic manner of speaking about the primary subject, but also actually refers to and truly describes some aspects of the structure of the primary domain. The consequence of this in scientific language is that the meaning of the original 'literal' language of the primary system is changed by introducing the metaphor (Hesse, 1983). This means that the distinction between observation and theoretical terms in science need be neither absolute nor universal. The observation-theoretical distinction can be drawn anew within each science as each new theory is incorporated into the existing body of scientific knowledge. Hesse (1983) claims that this comes about as a result of the gradual modification of the network of meanings in the observational terminology of an existing established theory (read: scientific description and explanation) by the (new) theoretical concepts which function as analogical extensions of this observational terminology.

In *The Cognitive Claims of Metaphor* (1983), Hesse parts company with Black (1979) concerning the issue of metaphorical dependency upon the relations of similarity and analogy between the primary and secondary systems, and argues that similarity and difference are irreducible primary relations, prior even to the application of the simplest predicate: they are *shown* and not *said* (1983:33). She claims the fact that the model does not say reality, but shows it, does not imply that there is a real structure of the world, of which science progressively exhibits more as it learns more of the natural environment. She argues that this is not a strong realism either, because science does not capture this structure in an isomorphism of true categories and true statements with the world. She says that 'all observation language is theory-laden' in the sense that all descriptive language carries along with it presupposed classifications of objects and properties in the world, and extra-empirical categorial frameworks that partially determine the acceptability or nonacceptability of descriptive statements within the language community. It follows that radical changes in theory are also radical changes in these presuppositions of language, and hence of the framework within which the observable natural world is described. Hesse (1983) describes her own position as that of 'moderate realism' which she elaborates as follows: The problem of explaining how models are significant in indicating the real, even though they cannot be taken as strongly realistic, is not due to their success, but to the fact that the reliability of models for prediction depends on analogical relations which they exhibit, rather than state, show rather than say, between diverse natural systems. Hesse finds it necessary to reject both the notion of the non-theory laden observation language and the idea that we can have an ideal scientific theory which explicitly states the true classification of the world. She rejects the isomorphism between language and the world, but claims that the model says as much as can be said within the linguistic resources available. She generalizes the thesis of the theory-ladenness of observation language to a thesis of the metaphoric character of all descriptive language. By this she means that extensions of meaning that occur in metaphor are only the more striking examples of something that is occurring all the time in natural language. In her final programmatic remarks on the cognitive claims of metaphor (1983; cf. also 1976:8), she states inter alia that all language, including 'ordinary descriptive language', is metaphorical in the sense that its use of general terms implies a normative classification of the multiplicity of things (1983:40). She rejects the so-called truth-conditional view of meaning which only attributes truth value to statements of a literal nature and argues for an 'intentional view of meaning' (1985/6) which allows for the possibility that metaphorical language could be qualified in as many ways as any other sentence could be. In the case of science and in the case of everyday action within the natural environment, the normative classification utilized in the network is constrained by criteria of prediction, test and self-correction. In a more general sense, Hesse (1983:41) claims, truth-value can be ascribed to some metaphoric statements within a social network of conventions, which may be as wide as a cultural vernacular, or internal to a particular ideology. The obvious question that follows from this argument is where and how these presupposed classifications of the categorial frameworks are anchored. She ultimately anchors the irreducible primary relations on which metaphors are based in 'social conventions' (1983:41).

Metaphors and Family Resemblances

In opposition to traditional views of metaphor which resort to some grounding in natural kinds or universals to which language is related, Hesse anchors metaphorical reference in the semantic network of the language. She rejects the Aristotelian theory of universals (which she calls the 'absolute theory') that views metaphor as the transposition of a name properly belonging to something else (ibid., 1983:28) and resorts to Wittgenstein's notion of irreducible 'family resemblances' and to the conventional nature of judgements concerning similarities and differences as an alternative. The 'absolute theory' would imply that the metaphor is in principle exhaustively paraphrasable by finding the proper word or words, therefore the net information gained by a metaphor, plus the restitution of proper meaning is nil – implying that metaphors have no cognitive but only decorative function.

The implication of the 'family resemblance' account of universals on the other hand, is that, strictly speaking, no meanings are univocal. The rules underlying the correct applications of meanings and underlying meaning relations must be sought. She concludes that the extensions of meaning that occur in metaphor are only the more striking examples of something occurring all the time in natural language and that if we wish to understand the process and judge its significance, it is to the elementary examples implied by the existence of family resemblance classes that we should first look. This is what she has in mind when she says: 'All language is metaphorical' (1983:40). She argues that the threat posed by various forms of relativism following the work of Kuhn and Feyerabend has led to the undermining of the belief in the reality of laws of nature and their corresponding universals. She writes:

> Radical revolutions of theoretical language call into question the possibility of reaching or even converging upon the ideal theory-language with its 'correct' classification of universals and hence laws, and if there is no convergence, may this not be because there are no ideal natural types (1985/6:37)?

She settles for a weaker form of realism (some would argue, for anti-realism) when she says that meanings of predicates in scientific language grow in dynamic interaction with culture and experience. Terms do not correspond to universals, Hesse claims. To the question, how we would know what we are talking about when confronted with the instability of meaning found in language in general she says:

> Members of a class are not identical realizations of an ontologically backed class-term, they are loosely grouped by relations of similarity and difference into fuzzy, overlapping and temporarily defined classes whose boundaries change with experience and cultural convention (1985/6:38; 1976:8).

In accepting Wittgenstein's view of 'family resemblances', she acknowledges that resemblances form, as it were, a chainlike structure through the class in such a way that there are relatively clear cases of objects falling within the class and relatively clear cases of those that do not. But this does not imply the existence of 'universals' (Hesse, 1985/6:38). Hesse claims that in spite of the instability of meaning in language and the absence of universals, communication is still possible in such a language, because much of the experience and convention is in common – these shared assumptions indeed constitute a language community. Learning and recognition of terms referring to certain states of affairs come about by a web of similarities and differences, '[...] the perception of which is also mostly common to our language group, because we share the same physiology and more or less the same experience and cultural expectations' (1985/6:39). There are irreducible perceptions of similarity and difference which are simply the 'function of our physiology and its commerce with the world' (1985/6:40). She notes: 'similarity' and 'difference' are not themselves universals, because there are no universals. Therefore, they must be judgements made on particular occasions by particular perceiving beings. This leads her in the direction of an anti-realist position, which she characterizes as 'moderate realism'.

Metaphors: A Realistic Portrayal of the World?

In her exposition of the family resemblances and the way in which the recognition and learning of these resemblances takes place Hesse (1985/6:40) stresses the fact that the classifications and categories thus formed, grow through appeals to notions such as 'the same experience ', 'shared assumptions', 'the same physiology', 'the same cultural expectations', 'irreducible perceptions' that are a function of 'our physiology and its commerce with the world'. In spite of her recognition of these 'objective realities' to which language is related she still opts for an anti-realist position. Her motivation is clearly the fact that our knowledge of these objective realities is limited, seldom definitive and always open to correction. By and large this is an assumption usually shared by anti-realist thinkers, yet it would be possible to subscribe to this same fact and still be sympathetic to a realist position, e.g., McMullin (1984:35). The source of this divergence must be sought elsewhere, it seems. In Hesse's case, the source of the divergence is related to her rejection of the 'absolute theory' of universals, which she claims to have evaded (or solved?) by the development of her own alternative view to the traditional view of universals, a version of Wittgenstein's family resemblance view.

Hesse (1983:28) argues that the family resemblances need not be supposed to be mysterious characteristics of the universe, except in the sense that people of a given culture, or sometimes all people, are found to agree in making judgements about them. To the question whether science is still possible in a world where there are no laws of nature (no 'natures' or 'universal laws'), she (1985/6:38) answers in the affirmative and suggests that progress in science is constituted by the growth of knowledge through a process in which, from language to language, certain local regularities remain approximately invariant and this stock of invariance grows and covers wider and wider local domains. Hesse opts for this view, because, she claims, this is actually a claim which is 'weaker on the demands it makes on ontology' than the so-called 'real universals view' (1985/6:38). Although Hesse places the emphasis on the dynamic semantic network of language, she acknowledges the fact that language grows in interaction with experience, but does not elaborate on the nature or structure of experience and the manner in which it conditions the formation of linguistic categories. A full-fledged theory of metaphor will at least have to provide some clues to the solution of this problem. Hesse's notion of context could provide an important starting point for the development of a theory of metaphor which recognizes that language is conditioned by more than only the linguistic constraints already present in the network of meanings of a language.

In her discussion concerning the manner in which the boundaries between the literal and the metaphorical change through the process of interaction, she argues for the recognition of the *context* and its role in the determination of what is literal and metaphorical. If metaphoricity is ultimately determined by context, how is context determined? On what basis do we differentiate between one context and another? Hesse's response would most probably be that context, in turn, is determined by social and cultural convention; that contexts too are fluid and continually changing and that the meta-problem (are contexts based on universals?) can be solved in exactly the same way as the problem concerning the reference of metaphors on the first order level: the constraints are built into the semantic network of a language developed in its commerce with the world. However, this approach still leaves us questioning exactly how this 'commerce with the world', experience and irreducible perceptions, is conditioned by the actual structure for the world. Can one have a theory of metaphor without it being based on some ontological framework of differences and similarities?

Metaphors and Universals

Hesse's argument concerning the grounds for similarities and differences and her rejection of the notion of universals is open to many questions: A categorical claim that there are no universals requires a stronger refutation than the arguments she advances, viz. the fluidity of semantic changes in a language which leads to continuous reformulating of the boundaries between the literal and the metaphorical, or 'proper' and 'improper' use of language and the absence of any real convergence in the direction of a true scientific account of the nature of the world. One salient point that she seems to overlook is the fact that it is exactly the presence of some sort of constancy which makes it possible to speak of change. Perhaps this is exactly what the traditional theory of universals was referring to. Why would it not be possible to recognize the dynamic changes in language and still hold on to a qualified notion of universals which in actual fact condition these dynamic changes? Admittedly, such an argument would require a different construal of the traditional notion of universals. One could share her rejection of Aristotle's 'absolute theory' of universals and its correlate theory of proper and improper or metaphorical language use. However, need this necessarily imply rejecting other accounts of what has traditionally been called the problem of universals? An Aristotelian theory misleads, making us think that objects can be parceled off into neat, sharply divided compartments, according as each possesses a certain characteristic, suggesting that the divisions between natural kinds are real and precise and discovered instead of recognizing that they are partly 'invented' by us (Woozley, 1976:80). This human contribution is well known in e.g., zoological classifications that change with the growth of new understanding. Moreover, it is clear that her acceptance and accommodation of Wittgenstein's notion of 'family resemblances' is already a modified version of the theory of universals; one which allows for the fact that meaning is contextually determined and that semantic change can come about through interaction between contexts.

To my mind the recognition of the fluidity of language and semantic change does not warrant the conclusion that there are no (contextual) conditions for predication and denomination, other than those imposed by the fluctuating semantic network of the language or mere social conventions. Semantic change and the possibility to recognize it are conditioned by more than semantic realities. Hesse seems to be aware of this fact when she points to the irreducible primary relations of difference and similarity. Moreover, these irreducible primary relations seem to be contextually determined. I believe that both these relations and their contextual qualifications are structural realities which are approximated by hypothetico-structural explanation (McMullin, 1978:139). In the approximation of these structures (also formulated as 'functional analogies' [Hart, 1984] or 'necessary metaphors' [Snell, 1953]) found in reality, metaphors play a constitutive role. This implies that the problem is not so much the notion of universals as such, but the traditional theory of universals. It therefore does not call for a rejection of the notion of universals, but a different construal or a reformulation of the problem of universals. This has the advantage of escaping the apparently relativistic and anti-realistic consequences which Hesse's position entails. Hart's (1984) formulation of the problem of the universals is helpful in this respect.

He argues that the universals should not be interpreted as universal entities that are instantiated and exemplified, but as structural conditions for the existence of empirical reality. These structural conditions are the bases of the irreducible primary relations such as similarity and difference in reality and also provide the bases for the possibility of linguistic predication and denomination. Each particular entity in our world has both individual and universal traits which are characteristic of all of its existence (Hart, 1984:18). In his ontology of universality Hart (1984) argues that the universality of the universals may point to something nomic, i.e. that all the various concepts used in talking about universals have in common that particulars must relate to them in a certain way if they are to be the particulars they are (Hart, 1984:35). Hart describes his position as follows:

> Both individuality and universality are real. They are also mutually irreducible and correlative. They are in fact traits of a relationship, namely, the relationship between the nomic conditions that hold universally for what exists and the empirical existents that individually and subjectively meet those conditions or are subject to them. And understanding what something is, is grasping in a concept what the conditions for its existence are (1984:83).

Concepts are therefore linguistic expressions of one's understanding of universal traits of individual and particular existents. We recognize these as belonging to certain categories or groups based on prescientific or scientific experiential knowledge. The conditions are not only characteristic of reality, but also provide the basis for the general modes of experience which warrant stability and the possibility of change.

Hart's position includes both a nominalist and a realist element: it rejects the notion of universals, yet acknowledges that whatever there is in the world does or must meet certain qualifications before it can be accepted as being of a certain kind (1984:19). Hart's view provides a viable alternative to the assumption shared by both the nominalist and the realist, viz. that there could be no objective justification for the application of a general term to its instances unless its instances have something in common, over and above their having in common that they are its instances.

Such a reformulation of the problem of the universals provides a different avenue to approach the problem of metaphorical models and scientific realism, one that is not necessarily incompatible with Hesse's appropriation of Wittgenstein's idea of family resemblances. The advantage of this approach is its appeal to *nomic conditions* which point to the fact that metaphorical reference is anchored in *more-than-semantic-reality*. Such an appeal is of course still confronted by the question concerning the proof of the existence of these nomic conditions. Hesse could respond that we still have no other access to these conditions than the linguistic conventions of our own culture. When confronted with the challenge to prove the existence of the nomic conditions, various answers are possible: One is a 'retreat to commitment', i.e. that one claims to believe that such conditions do exist. The other is an appeal to experience, which could be evidentially supported by scientific research. Obviously, these approaches need not be mutually exclusive. It is clear that the substantiation of a claim that such nomic conditions do structure cognitive and linguistic meaning would require evidence drawn from everyday experience of reality and from various disciplines. It would also require further development of the contours of a comprehensive philosophical ontology which could accommodate such empirical evidence. This has been dealt with briefly in the initial chapters of this book.

It is significant that various authors have suggested that such universal structures do exist; that they condition human and societal behaviour and that it would in principle be possible to construct a theory of these structural universals or a 'biogrammar' of the universal cultural acquisition device of humankind (cf. Harre, 1976:32; Tiger and Fox, 1974:17, 30). Cross cultural research in both psychology and anthropology have pointed to the existence of such traits, and in recent philosophical discussions Apel and others have pointed to the necessity of recognizing the existence of some sort of 'transcendentalia'. I believe these claims are also substantiated in the work of Lakoff and Johnson and others. Lakoff and Johnson (1980:17–19; 1999) argue that metaphorical concepts have experiential grounding. They write: '[...] no metaphor can ever be comprehended or even adequately represented independently of its experiential basis' (1980:19). One of their most significant conclusions is the statement that a 'basic domain of experience'

> [...] is a structured whole within our experience that is conceptualized as [...] an experiential gestalt. Such gestalts are experientially basic because they characterize structured wholes within recurrent human experiences. They represent coherent organizations of our experiences in terms of natural dimensions[...] Domains of experience that are organized as gestalts in terms of such natural dimensions seem to us to be natural kinds of experience (1980:117,118; 1981:19ff; 1999).

Lakoff & Johnson reject the idea of the grounding of the conceptual system within the framework of a so-called 'building block' theory in which all meaningful utterances are constructed from certain unanalyzable semantic units. Instead, they identify emergent categories and concepts that are best understood as experiential gestalts, which, although decomposable into other elements, are yet basic and irreducible in terms of grounding our conceptual system (1981:314–5). In their later work (1999) they speak of metaphor as the bodily based neural mechanism which ultimately also functions as the basis for human spirituality. Acknowledging the bodily and neural basis of metaphorical meaning is one thing. It is a very different matter when metaphorical meaning is moored to this basis. Mooring metaphorical meaning in its bodily basis is a bootstrap operation, because ultimately bodily configurations are in turn based upon and anchored in deeper ontic conditions that make human existence possible.

An adequate account of the realistic nature of theories based on metaphorical models requires more evidential support from various disciplines for the notion of structural conditions in the abovementioned sense. Recognising nomic conditions, is important and ultimately requires developing a philosophical ontology that articulates these conditions This would provide a framework determining semantic conflation (Spragens, 1973:41) in scientific concept formation – and prevent us from becoming the victims of a situation in theorizing when 'metaphors becomes myths' (Turbayne, 1970:28). Such a framework cannot be provided when metaphors are only anchored in the fluctuating semantic network of language.

NOTE

The initial draft of the published paper which forms the basis of this chapter was developed during a sabbatical at Notre Dame University in 1984, made possible by a Senior Researcher Grant of the Human Sciences Research Council of the Republic of South Africa. I would like to thank Ernan McMullin for helpful discussions and assistance during this time. This chapter under the title Metaphorical Models and Scientific Realism was published in PSA, 1986, volume I, pp 374 –383. Copyright by the Philosophy of Science Association. It has been reworked for this publication.

XI Embodiment and the Grammar of Creation

Real Certitudinal Anchors

Earlier it was argued that the methodology of metaphorical hermeneutics assumes an analogical structure to the world, human experience and cognition. This approach claims that metaphorical thought, images, language and models are able to access this structure of the world and that they mediate discovery and the opening up of the complex interrelated spheres of meaning of reality. This structure constitutes the 'moorings' of metaphorical meaning and reference. It is a realist position – but a qualified realism. Such a claim will no doubt be disputed by many involved in the realism-anti-realism debates, mainly because in the philosophical discussions realism is defined in diverse ways. (Delaney, 1985). The Rorty-Hesse, (1985/6) discussions concerning the reality of '[...] texts and types [...]' and '[...] lumps and laws [...]' and the Boyd (1979)-Kuhn (2000) discussion concerning the existence of natural kinds and/or accessibility of the 'joints' of the world, are cases in point. Whether this realist understanding of theoretical knowledge and the cognitive role of metaphor qualifies as a foundationalist approach is open to discussion. This is an important issue, but not one that I shall address at this stage.

Postmodernism questions any essentialist or foundationalist claims of knowledge and suspends the possibility of accessing reality outside of the interplay of texts and writing. The double-language view in which literal and metaphorical language is differentiated and juxtaposed often seeks the grounding or mooring of the meaning of the metaphorical concepts in some bedrock of literal meaning. Pan-metaphoricism on the other hand, argues that which is expressed can only be expressed metaphorically or is only accessible via metaphor. In general, these positions do not necessarily address the grounding or mooring of metaphors, nor do they necessarily have explicit positions on the cognitive status of metaphorical devices. Whether these metaphors have cognitive import, actually refer to states of affairs in reality and are constrained by the structure for reality, remains the contentious issue. The critical background issue in this debate is what Bernstein calls the 'Cartesian anxiety' which he identifies as the root or basis of the subjectivism-objectivism debate. He notes:

> The primary reason why the *agōn* between objectivists and relativists has become so intense today is the growing apprehension that there may be nothing – not God, reason, philosophy, science, or poetry – that answers to and satisfies our longing for ultimate constraints, for a stable and reliable rock upon which we can secure our thought and action (1983:18,19).

Foundationalism and its critics have developed a wide arsenal of responses to this quest – responses that each have a different answer to the question of the mooring or anchoring of metaphorical meaning. Answers range from reductionist positions that anchor metaphorical meaning in literal language to positions that seek the moorings of metaphorical meaning in social conventions and constructivist positions. Anti-realist positions often anchor meaning in some dimension of the subjective world, whereas realists argue that there is an objective and independent world out there which can be accessed. The realism espoused in this work argues for the real existence of conditions or structures for human existence, experience and reality. These structures not only condition cognition but also condition the empirical existence of all dimensions of reality.

Realism in everyday use of metaphor and scientific realism associated with the implementation of theory-constitutive metaphors in scientific theorizing have points of contact but are also vastly different. The difference in the cognitive import of metaphor is related to the widely divergent contexts in which they function. What is common to both everyday usage of metaphors and its use in science is the fact that it rests on human embodiment in the senses discussed in Part I. Even more fundamental is the fact that human relational embodiment in turn is anchored in the 'deeper grammar' of reality.

Both Bernstein (1983) and Slingerland (2004) grapple with the nature of this stable and reliable foundation or common core of human existence. Slingerland comes to the conclusion that there is a position between Enlightenment realism and postmodern anti-realism (2004:1–31) and argues for an '[...] embodied realism in which the commonalities of human bodily experience can serve as a basis for cross-cultural commensurability[...]'(2004:1). When demonstrating the commonalities between Confucian and Western theories of morality Slingerland states that '[...] both these theoretical conceptions grow out of and make use of a deeper metaphysical grammar that has its roots in common human embodied experience' (2004:16). He also calls this a common core. The crux of his argument is that these commonalities

> [...] are not reflections of some a priori order existing independently of humans, but arise out of the interaction of human bodies with a fairly stable physical world over the course of both evolutionary and personal time[...](ibid., 2004:17)

Against the anti-realists he would argue that there are structures of cognition common to all human beings regardless of their culture, language or a particular theory (2004:17). Slingerland therefore opts for embodied realism, a choice that I am willing to share but with a number of caveats.

The first caveat relates to a view of the creation order. I acknowledge the reality of the basis of metaphorical meaning in human embodiment, but also recognize the structural traits of all of non-human reality. The creation order exists both as condition for human and non-human reality (in that sense it is 'independent') but also takes shape in and through human and non-human response to these conditions. In the chapter dealing with the grounding of metaphorical meaning, we discussed the role of embodiment and the structures for creation that condition the existence of human embodiment. There is one important strand of this argument that still needs to be teased out. I argued that Lakoff and Johnson's suggestion of 'embodied realism' is a more acceptable and responsible realistic view than rationalistic realism, which hales from the Enlightenment. Yet, embodied realism as espoused by Lakoff and Johnson and their school of thought do not fully consider two dimensions. The first is that human embodiment is conditioned and structured by even deeper layers of reality, as grammar would structure a language. As argued earlier Hart does articulate a position that considers this: '[...] the thesis that the existing empirical world has an irreducible correlate in the order [for?] of the world' (1984:82, 83). He states that both individuality or particularity and universality are 'real', are mutually irreducible and correlative. This means that '[...] natural kinds, social order, norms for behaviour and laws are all real [...]'(ibid., 1984:82) as are the particular phenomena in reality that we designate as natural kinds, etc. Used in this sense the term 'realism' acquires a meaning different from the standard meaning found in most literature on the subject. With respect to realism in science theories developed based on metaphorical models and tested and corroborated in confrontation with empirical reality, provide a realistic approximation of reality.

This choice is made over and against the position of instrumentalism, which argues that metaphors are merely heuristic devices utilized for purposes of discovery. It is also a position, which is critical of anti-realism even though Mary Hesse's contributions to the discussions concerning realism have shaped my own understanding of the matter in many ways. However, Mary Hesse holds to an anti-realist or moderate realist position, mainly because of her rejection of the existence of universals, natural kinds and essences. I agree with her that the maintenance of a realist stance requires a rejection of the traditional (or absolute) theory of universals. Her network theory of meaning based on her appropriation of Wittgenstein's family resemblance notion is closer to a modified view of natural kinds and universals that I too find acceptable. But I do believe realism and scientific realism requires some construal or reformulation of a theory of universals and natural kinds in order to escape the potentially relativistic consequences which an anti-realist position entails. Such a theory of universals requires the recognition that the underlying classificatory system on which metaphorical reference is based represents more than conventional, sociologically determined semantic reality. It requires a positive recognition of the presence and knowability of God's presence in and through His creation order.

George Steiner states:

> [...]any coherent understanding of what language is and how language performs,[...]any coherent account of the capacity of human speech to communicate meaning and feeling is, in the final analysis, underwritten by the assumption of God's presence (1989:3).

He says that the wager on the meaning of meaning '[...]is a wager on transcendence' (1989:4, 214). Steiner's '[...]conjecture is that 'God' is, not because our grammar is outworn; but that grammar lives and generates worlds because there is a wager on God' (ibid.,

1989:4). And this wager is '[...] on the informing presence in the semantic markers which generates [...]' all kinds of works of art.

> They are re-enactments, reincarnations via spiritual and technical means of that which human questioning, solitude, inventiveness, apprehension of time and of death can intuit of the fiat of creation, out of which, inexplicably, have come the self and the world into which we are cast' (Steiner, 1989:215).

Elsewhere Steiner (2001) speaks of the 'grammar of creation'.

One could argue that 'knowing God' and 'knowing His order for His creation' are two different matters and that even if one were to concede the knowability of God through His presence in His creation, the order for creation still remains way beyond our theoretical grasp and can only be approximated. That being the case, I believe realism defined as the recognition of the existence of a creation order which we approximate inter alia with metaphorical models, is still a position preferable to anti-realism. A realism position constitutes a bulwark against relativism. In the inextricable correlation between the fiat or grammar of creation and our human articulation of this grammar via metaphor and analogy the mooring of metaphorical meaning becomes apparent.

Bibliography

Agassi, J. (2001) Chapter 9. In *Science in Flux. Boston Studies*, Vol 28, Dordrecht D. Reidel, pp. 208–239.

Altman, I. and Rogoff, B. (1987) World Views in Psychology: Trait, Interactional, Organism, and Transactional Perspectives. In Stokels, D. and Altman, I (eds.) *Handbook of Environmental Psychology*, New York, Wiley, pp. 7–40.

Anderson, R. (2006) The Crafting of Scientific Meaning and Identity. Exploring the Performative Dimensions of Michael Faraday's Texts. *Perspectives on Science*, Vol, 14, #1.

Arbib, M. and Hesse, M. (1986) *The Construction of Reality*. Cambridge, Cambridge University Press.

Bambrough, R. (1960–1961) Universals and Family Resemblances. *Proceedings of the Aristotelian Society for the Systematic Study of Philosophy Supplementary*, Vol. 61, pp. 207–22.

Bar-Hillel, Y. ed. (1972) Logic, Methodology and Philosophy of Science*; Proceedings of the Second International Congress for Logic, Methodology, and Philosophy of Science*. Amsterdam, North-Holland Pub. Co.

Bartholomew, C. (1995) Covenant and Creation: Covenant Overload or Covenantal Deconstruction. *Calvin Theological Journal* 30, pp. 11–33.

Beardsley, M. (1967) Metaphor. In Edwards, P. ed. *The Encyclopedia of Philosophy*, New York, Macmillan, pp. 284–89.

Berkson, W. (1974) *The Development of a World View from Faraday to Einstein*. London.

Bernstein, R. (1983) *Beyond Objectivism and Relativism: Science, Hermeneutics and Praxis*. Philadelphia, University of Pennsylvania Press.

Black, M. (1962) *Models and Metaphors: Studies in Language and Philosophy*. Ithaca, Cornell University Press.

Black, M. (1977) More About Metaphor. *Dialectica*, Vol. 31, no. 3–4, pp. 431–57.

Black, M. (1979) More About Metaphor. In Ortony, A. ed. *Metaphor and Thought*, Cambridge, Cambridge University Press, pp. 19–43.

Bois, J. (1978) *The Art of Awareness: A Textbook on General Semantics and Epistemics*. 3rd ed. Dubuque, Iowa Wm. C. Brown.

Boscovich, R. (1966) *A Theory of Natural Philosophy*. Put forward and explained by R. Boscovich. English ed. from the 1st Venetian text in 1763. Cambridge, MIT Press.

Botha, M. E. (1983) White and Black Civil Religion as Ideology. *Koers: Bulletin vir Christelike Wetenskap* Vol. 48, no. 4, pp.249–257.

Botha, M. E. (1986a) 'Bewitching' or Confusing Metaphysics? The Demarcation Between Science and Metaphysics According to Karl Popper. *Koers: Bulletin vir Christelike Wetenskap* Vol. 51, no. 4, pp. 430–40.

Botha, M. E. (1986b) Metaphorical Models and Scientific Realism. *Suid-Afrikaanse Tydskrif vir Wysbegeerte* Vol. 5, no. 3, pp. 83–87. Also published in Fine, A. & Machamer, P. *Proceedings of the 1986 Biennial Meeting of the Philosophy of Science Association,* Vol.1, Michigan East Lansing, pp. 374 – 383.

Botha, M. E. (1988a) Framework for a Taxonomy of Scientific Metaphor. *Philosophia Reformata: Orgaan van de Vereniging voor Calvinistische Wijsbegeerte* Vol. 53, no. 2, pp. 143–70.

Botha, M. E. (1988b) *Metaphorical* Models and Social Scientific Explanation. In *Essays in Social Theorizing,*. Pretoria Human Sciences Research Council, pp. 61–96.

Botha, M. E. (1993a) Michael Faraday en die Letterlike Lesing van die Twee Boeke van God: 'n Gevallestudie oor. *Die Verhouding van Geloof en Wetenskap. Koers*, 58(2), pp. 141–68 (*Michael Faraday and the Literal Reading of the Two Books of God*).

Botha, M. E. (1993b) The Mediating Role of Metaphor and Analogy in the Relationship Between Science and Religion. *Tydskrif vir Christelike Wetenskap*, jrg. 29, 4e kwartaal, pp. 60–73.

Botha, M. E. (1999) Spiritualiteit: Begrensing as Bevryding. In *Bewogen Realisme,* Kampen, Uitgeberij Kok, pp. 123 –30.

Botha, M. E. (2000a) Metaphor and Analogy Revisited. In Strauss, D. & Botting, M. eds. *Contemporary Reflections on the Philosophy of Herman Dooyeweerd: A Supplement to the Collected Works of Herman Dooyeweerd*, Lewiston, Edwin Mellen Press, pp. 49–82.

Botha, M. E. (2000b) Christian Scholarship for Reconciliation? The Free University of Amsterdam and Potchefstroom University for Christian Higher Education. In Van Vugt, W. & Daan Cloete, G. eds. *Race and Reconciliation in South Africa: a Multicultural Dialogue in Comparative Perspective*, Lanham, Md. Lexington Books, pp. 121–40.

Boyd, R. (1979) Metaphor and Theory Change: What is 'Metaphor' a Metaphor for? In Ortony, A. ed. *Metaphor and Thought* Cambridge, Cambridge University Press, pp. 356–408.

Bradie, M. (1980) Models, Metaphors and Scientific Realism. *Nature and System,* (2), pp. 3–20.

Bradie, M. (1984) The Metaphorical Character of Science. *Philosophia Naturalis* Vol. 21, no. 2–4, pp. 229–43.

Brandt, P. (2000) The Architecture of Semantic Domains: a Grounding Hypothesis in Cognitive Semiotics. *Revista Portuguesa de Humanidades*, 4(1 & 2), pp. 11–51.

Brooke, J. (1991) *Science and Religion: Some Historical Perspectives.* Cambridge, Cambridge University Press.

Brown, H. (1975) Paradigmatic Propositions. *American Philosophical Quarterly* Vol.12, pp. 85–90.

Brown, J. (1979) *Perception, Theory and Commitment: The New Philosophy of Science.* Chicago University of Chicago Press.

Brown, J. ed. (1984) *Scientific Rationality: The Sociological Turn.* Dordrecht D. Reidel.

Brown, R. (1977) *A Poetic for Sociology: Toward a Logic of Discovery for the Human Sciences.* Cambridge, Cambridge University Press.

Brown, R. (1994) Logics of Discovery as Narratives of Conversion: Rhetorics of Invention in Ethnography, Philosophy, and Astronomy. *Philosophy and Rhetoric* Vol. 27, no. 1 pp. 1–34.

Burrell, D. (1973) *Analogy and Philosophical Language.* New Haven, Yale University Press.

Burtt, E. (1954) *The Metaphysical Foundations of Modern Physical Science.* Rev. ed. Garden City, Doubleday.

Butcher, S. ed. (1911) *The Poetics.* 4th ed. London, Macmillan.

Cantor, G. (1985) Reading the Book of Nature: The Relation Between Faraday's Religion and His Science. In Gooding, D & Frank, A. James, eds. *Faraday Rediscovered: Essays on the Life and Work of Michael Faraday, 1791–1867*, New York, Stockton pp. 69–82.

Cantor, G., Gooding, D. & Frank A. (1991) *Faraday.* London Macmillan.

Cantor, G., Gooding, D. & Frank A, (1987) Weighing Light: The Role of Metaphor in Eighteenth-Century Optical Discourse. In Benjamin, A, Cantor, G & Christie, J. eds. *The Figural and the Literal: Problems of Language in the History of Science and Philosophy, 1630–1800.* Manchester, Manchester University Press., pp. 124–46.

Cantor, G., Gooding, D. & Frank A. (1991) *Michael Faraday: Sandemanian and Scientist: A Study of Science and Religion in the Nineteenth Century*. Basingstoke, UK Macmillan.

Cantor, G. (1996) Interpreting Michael Faraday as the 'Christian Philosopher': Some Problems Regarding Metaphysics. In van der Meer, J. ed. *Facets of Faith and Science. Vol. 1 Historiography and Modes of Interaction*. Lanham, MD, University Press of America, pp. 49–62.

Clines, D. (1996) Varieties of Indeterminacy. *Semeia* 71 pp.17–27.

Clouser, R. (1991) *The Myth of Religious Neutrality: An Essay on the Hidden Role of Religious Belief in Theories*. Notre Dame, University of Notre Dame Press.

Clouser, R. (2005) *The Myth of Religious Neutrality: An Essay on the Hidden Role of Religious Belief in Theories*. Rev. ed. Notre Dame, University of Notre Dame Press.

Collingwood, R. (1940) *An Essay on Metaphysics*. Oxford, Clarendon Press.

Collins, H. (1985) *Changing Order: Replication and Induction in Scientific Practice*. Chicago University of Chicago Press.

Davidson, D. (1978) What Metaphors Mean, *Critical Inquiry*, Vol. 5, No. 1, Special Issue on Metaphor (Autumn 1978), pp. 31–47.

Delaney, C. (1985) Presidential Address: Beyond Realism and Anti-realism. *Proceedings of the American Catholic Philosophical Association*. Notre Dame.

Dijksterhuis E. (1969) *The Mechanization of the World Picture*. London.

Dirven, R. (1994) Metaphor and Nation: Metaphors Afrikaners Live By, *Duisburger Arbeiten zur Sprach- und Kulturwissenschaft, Vol. 22*. Frankfurt am Main and New York, P. Lang.

Dirven, R., Hawkins, B. and Sandikcioglu, E. eds. (2001) *Language and Ideology*. Vol. 1, Theoretical and Cognitive Approaches, *Amsterdam Studies in the Theory and History of Linguistic Science. Series Iv, Current Issues in Linguistic Theory, Vol. 204–205*. Amsterdam and Philadelphia, J. Benjamins Pub. Co.

Dooyeweerd, H. (1953) *A New Critique of Theoretical Thought*. Freeman, D. & Young,W. trans. Vol. 1: *The Necessary Presuppositions of Philosophy*. Amsterdam, H. J. Paris.

Dooyeweerd, H. (1954) De Analogische Grondbegrippen der Vakwetenschappen en Hun Betrekking tot de Structuur van den Menschelijken Ervaringshorizon. *Mededeelingen der Koninklijke Nederlandsche Akademie van Wetenschappen, afd. Letterkunde, Nieuwe reeks* 17, no. 6 171–92. Knudsen:, R. trans. *The Analogical Concepts*.

Dooyeweerd, H. (1955) *A New Critique of Theoretical Thought.* Freeman, D & De Jongste, H. trans Vol. 2: *The General Theory of Modal Spheres.* Amsterdam, H. J. Paris.

Dooyeweerd, H. (1957) *A New Critique of Theoretical Thought.* Freeman, D & De Jongste, H. trans Vol. 3: *The Structures of Individuality of Temporal Reality.* Amsterdam, H. J. Paris.

Dooyeweerd, H. (1969) *A New Critique of Theoretical Thought.* Freeman, D & De Jongste, H. trans. Vol. 3: *The Structures of Individuality of Temporal Reality.* Philadelphia, Presbyterian and Reformed Pub. Co.

Edie, J. (1975) Identity and Metaphor: a Phenomenological Theory of Polysemy. *Journal of the British Society for Phenomenology* 6, no. 1 pp. 32–41.

Edwards, P. ed. (1967) *The Encyclopedia of Philosophy*, Vol. 5. New York, Macmillan.

Eubanks, P. (1999) The Story of Conceptual Metaphor: What Motivates Metaphoric Mappings? *Poetics Today* 20, no. 3, pp. 419–42.

Faraday, M. (1844a) A Speculation Touching Electric Conduction and the Nature of Matter. Experimental Researches in Electricity. In Hutchins, M. ed. (1952) *Great Books of the Western World*, Vol. 45. *Encyclopaedia Brittanica.* Chicago William Benton.

Faraday. M. (1844b) *Matter.* Manuscript dated 19 February 1844 in the Library of the Institution of Electrical Engineers, London.

Faraday, M. (1852) On the Physical Lines of Magnetic Force. Experimental Researches in Electricity. January 1844. In Hutchins, M. ed. (1952) *Great Books of the Western World*, Vol. 45. *Encyclopaedia Brittanica.* Chicago, William Benton.

Faraday, M. (1854) On Mental Education. In Faraday, M. *Experimental Researches in Chemistry and Physics.* Originally pub. 1859. Bicentennial edition. London Taylor and Francis, 1991.

Faraday, Michael. (1858) On the Conservation of Force. *Proceedings of the Royal Institution*, Feb. 27, 1857, Vol ii, p.352. In Faraday, M. *Experimental Researches in Chemistry and Physics.* Originally pub. 1859. Bicentennial edition. London Taylor and Francis, 1991.

Faraday, M. (1859) *Experimental Researches in Chemistry and Physics.* Originally pub. 1859. Bicentennial edition. London Taylor and Francis, 1991.

Fauconnier, G. (1994) *Mental Spaces: Aspects of Meaning Construction in Natural Language.* Cambridge, Cambridge University Press.

Funkenstein, A. (1986) *Theology and the Scientific Imagination From the Middle Ages to the Seventeenth Century.* Princeton, Princeton University Press.

Gadamer, H. (1975) *Truth and Method.* London Sheed and Ward.

Gadamer, H. (1989) *Truth and Method.,* 2nd rev. ed. London Sheed and Ward.
Gammie, J., & Perdue, L eds. (1990) *The Sage in Israel and the Ancient Near East.* Winona Lake Eisenbrauns.
Gaonkar, D. (1997) The Idea of Rhetoric in the Rhetoric of Science. In Gross, A. & Keith, W. eds. *Rhetorical Hermeneutics: Invention and Interpretation in the Age of Science.* Albany, State University of New York Press.
Gentner, Dedre. (1982) *Are Scientific Analogies Metaphors?* In *Metaphor: Problems and Perspectives,* Miall, D. ed. Brighton Harvester, pp. 106–32.
Gibson ,J. (1979) *The Ecological Approach to Visual Perception.* Houghton Mifflin.
Gill, J. (1996) *Wittgenstein and Metaphor.* New and rev. ed. Atlantic Highlands, NJ Humanities Press.
Glas, G. (2002) Churchland, Kandel, and Dooyeweerd on the Reducibility of Mind States. *Philosophia Reformata* (67), pp. 148–172.
Ginev, D. (1999) From a Strong Hermeneutics of Science to a Strong Rhetoric of Science. *Philosophy and Rhetoric* Vol. 32, no. 3, pp. 247–81.
Glas, J. (1811) *The Works of Mr. John Glas in Four Volumes.* Edinburgh, Alexander Donaldson.
Gooding, D. (1980). Metaphysic Versus Measurement: The Conversion and Conservation of Force in Faraday's Physics. *Annals of Science,* Vol. 37, pp. 1–29.
Gooding, D. (1982) Empiricism in Practice: Teleology, Economy and Observation in Faraday's Physics. *Isis,* Vol. 73, pp. 46–67.
Goudzwaard, R. & De Lange, H. (1991) *Beyond Poverty and Affluence: Towards a Canadian Economy of Care.* Toronto, University of Toronto Press.
Grady, J., Oakley, T. & Coulson, S. (1999) Blending and Metaphor. In Gibbs, R. Jr., & Steen, G. *Metaphor in Cognitive Linguistics: Selected Papers from the Fifth International Cognitive Linguistics Conference,* Amsterdam, July 1997. Amsterdam & Philadelphia, John Benjamin.
Griffioen, S. (1997) Cultuuromslag Aan De V.U. In Verhoogt, J., Griffioen, S., & Fernhout, R. eds. *Vinden en Zoeken: Het Bijzondere van de Vrije Universiteit,* Kampen: Kok, pp. 61–81.
Gross, A., & Keith, W. eds. (1997) *Rhetorical Hermeneutics: Invention and Interpretation in the Age of Science.* Albany, State University of New York Press.
Haack, S. (1987/8) Surprising Noises: Rorty and on Metaphor. In Edgington, D. ed. The Aristotelian Society *Proceedings of The Aristotelian Society* Vol. LXXXVIII, London, pp. 293–301.
Hacking, I. (1983) *Representing and Intervening: Introductory Topics in the Philosophy of Natural Science.* Cambridge, Cambridge University Press.

Harré, R. (1976) The Constructive Role of Models. In Collins, L. ed. *The Use of Models in the Sciences.* London Tavistock, pp. 16–43.
Harré, R. (1978) *The Philosophies of Science: An Introductory Survey.* Oxford, Oxford University Press.
Harré, R. and Martin, J. (1982) Metaphor in Science. In Miall, D. ed. *Metaphor: Problems and Perspectives.* Sussex, The Harvester Press, pp. 89–105.
Harrington, A. (1991) Interwar 'German' Psychobiology: Between Nationalism and the Irrational. *Science in Context* Vol. 4, no. 2 pp. 429–47.
Harrington, A. (1995) Metaphoric Connections: Holistic Science in the Shadow of the Third Reich. *Social Research* Vol. 62, no. 2, pp. 357–85.
Harrington, A. (1996) *Reenchanted Science: Holism in German Culture from Wilhelm II to Hitler.* Princeton, NJ Princeton University Press.
Hart, H. (1984) *Understanding Our World: An Integral Ontology.* Lanham, University Press of America.
Hart, H. (1995) Creation Order in Our Philosophical Tradition: Critique and Refinement. In *An Ethos of Compassion and the Integrity of Creation*, edited by Walsh, B., Hart, H., & VanderVennen, R. Lanham University Press of America, pp. 67–96.
Haynes, F. (1975) Metaphor as Interactive. *Educational Theory* Vol. 25, no. 3 pp. 272–77.
Heelan, P. (1972a) Hermeneutics of Experimental Science in the Context of the Life-world. *Philosophia Mathematica* 9, pp. 101–44.
Heelan, P. (1972b) Towards a Hermeneutic of Natural Science. *Main Currents in Modern Thought* Vol. 28, no. 3, pp. 85–93.
Heelan, P. (1972c) Nature and Its Transformations. *Theological Faculties of the Society of Jesus Theological Studies.* Vol. 33 (3) pp. 486–502.
Heelan, P. (1983a) Natural Science as a Hermeneutic of Instrumentation. *Philosophy of Science* Vol. 50, pp. 181–204.
Heelan, P. (1983b) Perception as a Hermeneutical Act. *Review of Metaphysics* Vol. 37, pp. 61–75.
Heidegger, M. (1962) *Being and Time.* Macquarrie, J. & Robinson, E. trans. New York, Harper and Row.
Heimann, P. (1971) Faraday's Theories of Matter and Electricity. *The British Journal for the History of Science,* Vol. 5:19, pp. 235–257.
Hekman, S. (1984) Action as a Text: Gadamer's Hermeneutics and the Social Scientific Analysis of Action. *Journal for the Theory of Social Behaviour* Vol. 14, no. 3, pp. 333–54.

Helman, D. ed. (1988) *Analogical Reasoning: Perspectives of Artificial Intelligence, Cognitive Science, and Philosophy*. Dordrecht, Kluwer Academic Publishers.

Hempel, C. (1965) *Aspects of Scientific Explanation, and Other Essays in the Philosophy of Science*. New York and London, Free Press and Macmillan.

Hesse, M. (1955) Action at a Distance in Classical Physics. *Isis,* Vol. 46 (4), no. 146, December, pp. 337–353.

Hesse, M. (1961) *Forces and Fields. The Concept of Action at a Distance in the History of Physics*. London, Thomas Nelson.

Hesse, M. (1966) *Models and Analogies in Science*. Notre Dame, Notre Dame University Press.

Hesse, M. (1970) Is There an Independent Observation Language? In Colodny, R. ed. *The Nature and Function of Scientific Theories: Essays in Contemporary Science and Philosophy*, pp. 35–77. Pittsburgh, University of Pittsburgh Press.

Hesse, M. (1972) The Explanatory Function of Metaphor. In Bar-Hillel, Y. ed. *Logic, Methodology and Philosophy of Science; Proceedings of the Second International Congress for Logic, Methodology, and Philosophy of Science*, Amsterdam, North-Holland Pub. Co., pp. 249–59.

Hesse, M. (1974) *The Structure of Scientific Inference*. London, Macmillan.

Hesse, M. (1975) Models of Method in the Natural and Social Sciences. *Methodology and Science* pp.163–178.

Hesse, M. (1976) *Models Versus Paradigms in the Natural Sciences*. In Collins, L. ed. *The Use of Models in the Sciences*. London, Tavistock, pp. 1–15.

Hesse, M. (1980) *Revolutions and Reconstructions in the Philosophy of Science, Harvester Studies in Philosophy; No. 17*. Brighton, Sussex Harvester Press.

Hesse, M. (1983) The Cognitive Claims of Metaphor. In van Noppen, J. ed. *Metaphor and Religion*, Brussel: Vrije Universiteit Brussels, pp. 27–45.

Hesse, M. (1985–86) Texts Without Types and Lumps Without Laws. *New Literary History* 17, pp. 31–48.

Hesse, M. (1986) Changing Concepts and Stable Order. Review of H.M. Collins, 1985. *Social Studies of Science* 16, pp. 714–26.

Hesse, M. (1987) Unfamiliar Noises II: Tropical Talk: The Myth of the Literal. *Proceedings of the Aristotelian Society for the Systematic Study of Philosophy Supplementary Vol.* 61, pp. 297–311.

Hesse, M. (1988a) Theories, Family Resemblances and Analogy. In Helman, D. ed. *Analogical Reasoning: Perspectives of Artificial Intelligence, Cognitive Science, and Philosophy*, pp. 317–40. Dordrecht, Kluwer Academic Publishers.

Hesse, M. (1988b) Socializing Epistemology. In McMullin, E. ed. *Construction and Constraint: The Shaping of Scientific Rationality*, Notre Dame, University of Notre Dame Press, pp. 97–122.

Hirsch , E.D. Jr. (1967) *Validity in Interpretation.* New Haven, Yale University Press.

Hoffman, R. (1980) Metaphor in Science. In Honeck, R. & Hoffman, R. eds. *Cognition and Figurative Language*, Hillsdale, NJ Lawrence Erlbaum Associates, pp. 393–403.

Jammer, M. (1957) *Concepts of Force. A Study in the Foundations of Dynamics.* New York, Harper Brothers.

Jennings, W. (1997) Baptizing a Social Reading: Theology, Hermeneutics and Postmodernity. In Lundin, R. ed. *Disciplining Hermeneutics: Interpretation in Christian Perspective*, Grand Rapids, W. B. Eerdmans, pp.117–27.

Johnson, M. (1980) A Philosophical Perspective on the Problems of Metaphor. In Honeck, R. & Hoffman, R. eds. *Cognition and Figurative Language*, Hillsdale, NJ, Lawrence Erlbaum Associates, pp. 46–67.

Johnson, M. ed. (1981) *Philosophical Perspectives on Metaphor.* Minneapolis, University of Minnesota Press.

Johnson, M. (1987) *The Body in the Mind: The Bodily Basis of Meaning, Imagination, and Reason.* Chicago, University of Chicago Press.

Johnson, M. (1988) Some Constraints on Embodied Analogical Understanding. In Helman, D. ed. *Analogical Reasoning: Perspectives of Artificial Intelligence, Cognitive Science, and Philosophy*, Dordrecht, Kluwer Academic Publishers, pp. 25–40.

Johnson, M. (1989) Image-schematic Bases of Meaning. *Semiotic Inquiry* Vol. 9, no. 1–3, pp. 109–18.

Johnson, M. (1991) Knowing Through Body. *Philosophical Psychology* Vol. 4, no. 1, pp. 3–18.

Johnson, M. (1993a) Conceptual Metaphor and Embodied Structures of Meaning: A Reply to Kennedy and Vervaeke. *Philosophical Psychology*, 6(4), pp. 413–422.

Johnson, M. (1993b) *Moral Imagination: Implications of Cognitive Science for Ethics.* Chicago, University of Chicago Press.

Kalsbeek, L. (2002) *Contours of a Christian Philosophy: An Introduction to Herman Dooyeweerd's Thought.* Lewiston, The Edwin Mellon Press.

Kant, I. (1786) *Philosophy of Material Nature.* Ellington, W, trans. Indianapolis, Hacket Publishing Co.1985.

Kennedy, J., & Vervaeke, J. (1993) Metaphor and Knowledge Attained Via the Body. *Philosophical Psychology* Vol. 6, no. 4, pp. 407–12.

Kittay, E. & Lehrer, A. (1981) Semantic Fields and the Structure of Metaphor. *Studies in Language* Vol. 5, no. 1, pp. 31–63.

Kittay, E. (1987) *Metaphor: Its Cognitive Force and Linguistic Structure.* Oxford, Clarendon Press.

Kittay, E. (1995) A Reply to Donald Davidson's 'What Metaphors Mean' or Metaphors as Rearranging the Furniture of the Mind. In Radman, Z. ed. *From a Metaphorical Point of View: A Multidisciplinary Approach to the Cognitive Content of Metaphor*, Berlin and New York, W. de Gruyter, pp. 73–116.

Korzybski, A. (1933) *Science and Sanity: An Introduction to Non-Aristotelian Systems and General Semantics.* Lancaster, Pa; New York, The International Non-Aristotelian Library Publishing Company/The Science Press Printing Company, distributors.

Kövecses, Z. (2002) *Metaphor: A Practical Introduction.* New York, Oxford University Press.

Kövecses, Z. (2005) *Metaphor in Culture: Universality and Variation.* Cambridge, Cambridge University Press.

Kuhn, T. (1973) *The Structure of Scientific Revolutions.* 2nd enlarged ed. Chicago Chicago University Press.

Kuhn, T. (1974) Second Thoughts on Paradigms. In Suppe, F. ed. *The Structure of Scientific Theories*, Urbana, University of Illinois Press, pp. 459–499.

Kuhn, T. (1977) *The Essential Tension. Objectivity, Value Judgement and Theory Choice.* Chicago, University of Chicago.

Kuhn, T. (1979) Metaphor in Science. In Ortony, A. ed. *Metaphor and Thought*, Cambridge, Cambridge University Press, pp. 533–542.

Kuhn, T. (1989) Possible Worlds in History of Science. In Allén, S. ed. *Possible Worlds in Humanities, Arts and Sciences: Proceedings of Nobel Symposium 65*, Berlin Walter de Gruyter, pp.7–22.

Kuhn, T. (1993) Metaphor in Science. In Ortony, Andrew (ed). *Metaphor and Thought* 2nd ed. Cambridge, Cambridge University Press.

Kuhn, T. (2000) The *Road Since Structure: Philosophical Essays, 1970–1993, with an Autobiographical Interview.* Chicago, University of Chicago Press.

Kvale, S. (1995) Themes of Postmodernity. In Anderson, W. ed. *The Truth About the Truth: De-confusing and Re-constructing the Postmodern World*, New York, Putnam, pp. 18–25.

Lakatos, I. (1974) Falsification and the Methodology of Scientific Research Programmes. In Musgrave, A and Lakatos, I. eds. *Criticism and the Growth of Knowledge*, Cambridge, Cambridge University Press.

Lakoff, G. (1986) The Meanings of Literal. *Metaphor and Symbolic Activity* vol.1, no. 4, pp. 291–96.

Lakoff, G. (1987) *Women, Fire and Dangerous Things: What Categories Reveal About the Mind.* Chicago, University of Chicago Press.

Lakoff, G. (1993) The Contemporary Theory of Metaphor. In Ortony, A. ed. *Metaphor and Thought*. Cambridge, Cambridge University Press pp.202–251.

Lakoff, G. & Johnson, M. (1980) *Metaphors We Live By*. Chicago, University of Chicago Press.

Lakoff, G. & Johnson, M. (1981a) Conceptual Metaphor in Everyday Language. In Johnson, M. ed. *Philosophical Perspectives on Metaphor*, Minneapolis, University of Minnesota Press, pp. 286–328.

Lakoff, G. & Johnson, M. (1981b) The Metaphorical Structure of the Human Conceptual System. In Norman, D. ed. *Perspectives on Cognitive Science*, Norwood, NJ, Ablex Publishing Co., pp.193–206.

Lakoff, G. & Johnson, M. (1988) Cognitive Semantics. In Umberto, E. Santambrogio, M & Violi, P. *Meaning and Mental Representation.* Bloomington Indiana, University Press, pp. 119–154.

Lakoff, G. & Johnson, M. (1999) *Philosophy in the Flesh: the Embodied Mind and Its Challenge to Western Thought.* New York, Basic Books.

Lakoff, G. & Turner, M (1989) *More Than Cool Reason: A Field Guide to Poetic Metaphor.* Chicago, University of Chicago Press.

Laszlo, E. (1972) *Introduction to Systems Philosophy: Toward a New Paradigm of Contemporary Thought.* New York, Harper and Row.

Leddy, T. (1995) Metaphor and Metaphysics. *Metaphor and Symbolic Activity* vol.10, no. 3, pp. 205–22.

Leplin, J. ed. (1984) *Scientific Realism.* Berkeley, University of California Press.

Leplin, J. ed. (1988) Is Essentialism Unscientific? *Philosophy of Science* vol. 55, no. 4, pp. 493–510.

Levere, T. (1968) Faraday, Matter and Natural Theology – Reflections on an Unpublished Manuscript. *British Journal for the History of Science*, 4, 95–107.

Levere, T. (1971) *Affinity and Matter. Elements of Chemical Philosophy 1800–1865.* Oxford, Clarendon Press.

Levere, T. (1981) *Poetry Realized in Nature. Samuel Taylor Coleridge and Early Nineteenth-century Science.* Cambridge, Cambridge University Press.

Lewin, K. (1959) *Field Theory in Social Science: Selected Theoretical Papers*. London, Tavistock.

Little,W., Fowler, H. & Coulson, J. (1975) *The Shorter Oxford English Dictionary on Historical Principles*. 3rd ed. Vol. 2. Oxford, Clarendon Press.

Luckman, J. (1986) *R. G. Collingwood's Doctrine of Absolute Presuppositions and its Bearing on the Problem of Historical Understanding*. Ph.D., McMaster University, Hamilton, Ontario.

Maasen, S. (2000) Metaphors in the Social Sciences. In Hallyn, F. ed. *Metaphor and Analogy in the Sciences*, Dordrecht, Kluwer Academic Publishers, pp. 199–244.

MacCormac, E. (1985) *A Cognitive Theory of Metaphor*. Cambridge, MIT Press.

Malachowski, A. ed. (1990) *Reading Rorty: Critical Responses to Philosophy and the Mirror of Nature (and Beyond)*, Oxford, Blackwell.

Martin, T. ed. (1932–36) *Faraday's Diary. Being the Various Philosophical Notes of Experimental Investigation Made by Michael Faraday, DCL, FRS, During the Years 1820–1862 and Bequeathed by Him to the Royal Institution of Great Britain*. 7 Vol. plus index. London. Cited as Faraday's Diary.

McFague, S. (1985) *Metaphorical Theology: Models of God in Religious Language*. Philadelphia, Fortress Press.

McFague, S. (1987) *Models of God: Theology for an Ecological, Nuclear Age*. Philadelphia, Fortress Press.

McGrath, A. (1998) *The Foundations of Dialogue in Science and Religion*. Oxford, Blackwell.

McMullin, E. (1968) What Do Physical Models Tell Us? In van Rootselaar, B. & Staal, J. eds. *Logic, Methodology and Philosophy of Science III: Proceedings of the Third International Congress for Logic, Methodology and Philosophy of Science, Amsterdam 1967*, Amsterdam, North-Holland Pub. Co., pp. 385–96.

McMullin, E. (1974) Models and Analogies in Science. In *New Catholic Encyclopedia*, Washington DC, Publisher's Guild, pp. 295–99.

McMullin, E. (1978) Structural Explanation. *American Philosophical Quarterly* 15, pp. 139–47.

McMullin, E. (1984) A Case for Scientific Realism. In Leplin, J. ed. *Scientific Realism*, Berkeley, University of California Press, pp. 8–40.

Mellor, D. (1968) Models and Analogies in Science: Duhem versus Campbell? *Isis* 59, pp. 282–90.

Miller, J. (1969) Science and Religion: Their Logical Similarity. *Religious Studies* 5, pp. 49–68.

Mouton, J, Joubert, D., Botha, E., Retief, A., Mos, L. (1988) *Essays in Social Theorizing*. Pretoria, Human Sciences Research Council, Research Report Series, no. 9.

Nersessian, N. (1984a) *Science and Philosophy. Faraday to Einstein: Constructing Meaning in Scientific Theories*. Dordrecht, Marthinus Nijhoff Publishers.

Nersessian, N. (1984b) Aether/Or: The Creation of Scientific Concepts. *Studies in the History and Philosophy of Science*, Vol. 15, no. 3, pp.175–212.

Nersessian, N. (1987) A Cognitive-historical Approach to Meaning in Scientific Theories. In Nersessian, N. ed *The Process of Science: Contemporary Philosophical Aapproaches to Understanding Scientific Practice* Dordrecht, Martinus Nijhoff, pp. 167–77.

Nersessian, N. (1991a) The Cognitive Sciences and the History of Science. In *Critical Problems and Research Frontiers in History of Science and History of Technology*, 30 October – 3 November 1991. Madison, Wisconson, pp. 92–115.

Nersessian, N. (1991b) Discussion: The Method to Meaning: a Reply to Leplin. *Philosophy of Science* 58, no. 4, pp. 678–86.

Nersessian, N. (1992) How Do Scientists Think? Capturing the Dynamics of Conceptual Change in Science. In Giere, R ed. *Cognitive Models of Science*. Minneapolis, University of Minnesota Press, pp. 5–22.

Neurath, O. (1973) *Empiricism and Sociology*. Foulkes, P. & Neurath, M. trans. Dordrecht, Reidel.

Olthuis, J. (1976) The Word of God and Biblical Authority, Wetenskaplike Bydraes van die PU vir CHO Reeks F. *Instituut Vir die Bevordering van Calvinisme, Studiestuk Nr. 101*. Potchefstroom: PU vir CHO.

Olthuis, J. (1985) "On Worldviews." *Christian Scholar's Review* 14, no. 2, pp. 153–64.

Olthuis, J. (1987) Proposal for a Hermeneutics of Ultimacy. In Olthuis, J. ed. *A Hermeneutics of Ultimacy: Peril or Promise?* Lanham, University Press of America.

Olthuis, J., Bloesch, D., Pinnock, C. & Sheppard, G. (1987) *A Hermeneutics of Ultimacy: Peril or Promise?* Lanham, University Press of America.

Ortony, A. (1975) Why Metaphors Are Necessary and Not Just Nice. *Educational Theory* 25, pp. 45–53.

Ortony, A. ed. (1979) *Metaphor and Thought*. Cambridge, Cambridge University Press.

Ortony, A. ed. (1993) *Metaphor and Thought*. 2nd ed. Cambridge, Cambridge University Press.

Pepper, S. (1942) *World Hypotheses: A Study in Evidence*. Berkeley, University of California Press.

Pepper, S. (1970) *World Hypotheses: A Study in Evidence.* 6th ed. Berkeley, University of California Press.

Pepper, S. (1973) Metaphor in Philosophy. In Wiener, P. ed. *Dictionary of the History of Ideas: Studies of Selected Pivotal Ideas,* New York, Scribner, pp. 196–201. Reprint in *The Journal of Mind and Behavior* 3, no. 3 (1982), pp. 197–206.

Perdue, L. (1994) *Wisdom and Creation: The Theology of Wisdom Literature.* Nashville, Abingdon Press.

Polanyi, M. (1974) *Personal Knowledge. Towards a Post-critical Philosophy.* Chicago, University of Chicago Press.

Polkinghorne, J. (1991) *Reason and Reality: The Relationship Between Science and Theology.* Philadelphia, Trinity Press International.

Popper, K. (1965) *Conjectures and Refutations: The Growth of Scientific Knowledge.* New York, Harper and Row.

Postman, N. (1996) *The End of Education: Redefining the Value of School.* New York, Vintage.

Poythress, V. (1987) *Symphonic Theology: The Validity of Multiple Perspectives in Theology.* Grand Rapids, Zondervan,.

Quine, W. (1975) On Empirically Equivalent Systems of the World. *Erkenntnis* 9, pp. 313–28.

Richards, I. (1936) *The Philosophy of Rhetoric.* New York and London, Oxford University Press.

Reddy, M. (1993) The Conduit Metaphor: A Case of Frame Conflict in Our Language About Language. In Ortony, A. *Metaphor and Thought.* 2nd ed. Cambridge, Cambridge University Press, pp. 164–201.

Ricœur, P. (1971) The Model of the Text. *Social Research,* 38, pp. 529–55.

Ricœur, P. (1974/5) Metaphor and the Main Problem of Hermeneutics. *New Literary History.* 6: 9–110.

Ricœur, P. (1976a) *Interpretation Theory: Discourse and the Surplus of Meaning.* Fort Worth, Texas Christian University Press.

Ricœur, P. (1976b) History and Hermeneutics. *Journal of Philosophy* vol. 73, pp. 683–95.

Ricœur, P. (1977a) The Model of the Text: Meaningful Action Considered as a Text. In Dallmayr, F. & McCarthy, T. ed. *Understanding and Social Inquiry,* Notre Dame, University of Notre Dame Press, pp.316–34.

Ricœur, P. (1977b) *The Rule of Metaphor: Multi-disciplinary Studies in the Creation of Meaning in Language.* Toronto, University of Toronto Press.

Ricœur, P. (1978) The Metaphorical Process as Cognition, Imagination, and Feeling. *Critical Inquiry* vol. 5, no. 1, pp.143–59.

Ricœur, P. (1980) The Bible and the Imagination. In Betz, H. ed. *The Bible as a Document of the University*, Chico, Calif., Scholars Press, pp. 49–75.
Rogers, C. (1981) The Formative Tendency. In Royce, J. & Mos, L. eds. *Humanistic Psychology: Concepts and Criticisms,* New York, Plenum Press, pp. vii–x.
Rohrer, T. (2001) Pragmatism, Ideology and Embodiment: William James and the Philosophical Foundations of Cognitive Linguistics. In Dirven, R., Hawkins, B. & Sandikcioglu, E. *Language and Ideology*, Vol. 1 *Theoretical and Cognitive Approaches*, Amsterdam and Philadelphia, J. Benjamins Pub. Co., pp. 49–82.
Rohrer, T. (2005) Embodiment and Experientialism. In Geerarts, D. & Cuyckens, H., eds. *The Handbook of Cognitive Linguistics.* Oxford, Oxford University Press, pp. 49–91.
Rorty, R. (1981) *Philosophy and the Mirror of Nature.* Princeton Princeton University Press.
Rorty, R. (1985/6) Texts and Lumps (of Matter). *New Literary History*, 17, pp. 31–48.
Rorty, R. (1987) Unfamiliar Noises: Hesse and Davidson on Metaphor. *Proceedings of the Aristotelian Society for the Systematic Study of Philosophy Supplementary* Volume 61, pp. 283–311.
Rorty, R. (1989) *Contingency, Irony and Solidarity.* Cambridge, Cambridge University Press.
Rosch, E. (1975a) Cognitive Reference Points. *Cognitive Psychology*, 7, pp. 532–547.
Rosch, E. & Mervis, C. (1975b) Family Resemblances: Studies in the Internal Structure of Categories. *Cognitive Psychology 7,* pp.573–605.
Rothbart, D. (1984) The Semantics of Metaphor and the Structure of Science. *Philosophy of Science* 51, pp.595-615.
Rothbart, D. (1988) Analogical Information Processing Within Scientific Metaphors. In Helman, D. ed. *Analogical Reasoning: Perspectives of Artificial Intelligence, Cognitive Science, and Philosophy*, Dordrecht, Kluwer Academic Publishers, pp. 377–99.
Royce, J. & Mos, L. eds. (1981) *Humanistic Psychology: Concepts and Criticisms.* New York, Plenum Press.
Rudwick, M. (1981) Senses of the Natural World and Senses of God. Another Look at the Historical Relation of Science and Religion. In Peacocke, A. ed. *The Sciences and Theology in the Twentieth Century.* Stocksfield, Oriel Press.
Scheffler, I. (1979) *Beyond the Letter: A Philosophical Inquiry Into Ambiguity, Agueness and Metaphor in Language.* London, Routledge & Kegan Paul.
Schön, D. (1963a) *Displacement of Concepts.* London, Tavistock.

Schön, D. (1963b) *Invention and the Evolution of Ideas*. Frome, Butler and Tanner.
Schön, D. (1979) Generative Metaphor: A Perspective on Problem-setting in Social Policy. In Ortony, A. ed. *Metaphor and Thought*, Cambridge, Cambridge University Press, pp. 254–83.
Searle, J. (1988) *Intentionality: An Essay in the Philosophy of Mind*. Cambridge, Cambridge University Press.
Seerveld, C. (1980) *Rainbows for a Fallen World. Aesthetic Task and Artistic Life*. Toronto, Tuppence Press.
Shanon, B. (1988) Semantic Representation of Meaning: A Critique. *Psychological Bulletin*, vol. 104, no. 1, pp. 70–83.
Shanon, B. (1992) Metaphor: From Fixedness and Selection to Differentiation and Creation. In Shenn, Y ed. Aspects of Metaphor Comprehension. Special issue. *Poetics Today* vol.. 13, no. 4, pp. 659–85.
Shanon, B. (1993) *The Representational and the Presentational: An Essay on Cognition and the Study of Mind*. New York, Harvester Wheatsheaf.
Sinnema, D. (1975) *The Uniqueness of the Language of Faith with Special Reference to the Language of Scripture*. Toronto, Institute for Christian Studies.
Slingerland, E. (2004) Conceptual Metaphor Theory as Methodology for Comparative Religion. *Journal of the American Academy of Religion*, Vol. 74, no1, pp. 1–31, March.
Smith, D. & Carvill, B. (2000) *The Gift of the Stranger: Faith, Hospitality and Foreign Language Learning*. Grand Rapids, *Eerdmans*.
Snell, B. (1953) *The Discovery of the Mind: The Greek Origins of European Thought*. Oxford, Blackwell.
Soskice, J. (1985) *Metaphor and Religious Language*. Oxford, Clarendon Press.
Soskice, J. (1987) Theological Realism. In Abraham, W. & Holtzer, S. ed. *The Rationality of Religious Belief: Essays in Honour of Basil Mitchell*, Oxford Clarendon Press, pp. 105–120.
Spencer, J. (1967/68) Boscovich's Theory and Its Relation to Faraday's Researches: An Analytic Approach. In Truesdell, C. ed. *Archive for History of Exact Sciences*, Vol. 4 Berlin, Springer-Verlag.
Spragens, T. (1973) *The Dilemma of Contemporary Political Theory: Toward a Postbehavioral Science of Politics*. New York, Dunellen.
Spykman, G. (1992) *Reformational Theology: A New Paradigm for Doing Dogmatics*. Grand Rapids, Eerdmans.
Stafleu, M. (1978) The Mathematical and Technical Opening Up of a Field of Science. *Philosophia Reformata*, Vol. 43(1), pp. 18–37.

Stafleu, M. (1979) The Isolation of a Field of Science. *Philosophia Reformata,*, Vol. 44(1), pp. 15–27.
Stafleu, M. (1980) *Time and Again: A Systematic Analysis of the Foundations of Physics.* Toronto, Wedge.
Stafleu, M. (1982) Theories as Logically Qualified Artefacts (II). *Philosophia Reformata: Orgaan van de Vereniging voor Calvinistische Wijsbegeerte* vol. 47, no. 1, pp. 20–40.
Stafleu, M. (1987) *Theories at Work: On the Structure and Functioning of Theories in Science, in Particular During the Copernican Revolution.* Lanham, University Press of America.
Stanford, W. (1936) *Greek Metaphor: Studies in Theory and Practice.* Oxford, B. Blackwell.
Stein, M. (1963) The Poetic Metaphors of Sociology. In Stein, M. & Vidich, A. eds. *Sociology on Trial*, Englewood Cliffs, N.J., Prentice-Hall, pp. 173–182.
Steiner, G. (1989) *Real Presences.* Chicago University of Chicago Press.
Steiner, G. (2001) *Grammar of Creation.* London, Faber & Faber.
Stoker, H. (1967) *Oorsprong en Rigting.* Vol. 1. Kaapstad Tafelberg.
Strauss, D. (1973) *Begrip en Idee.* Assen, Gorcum.
Strauss, D. (1984) An Analysis of the Structure of Analysis. *Philosophia Reformata: Orgaan van de Vereniging voor Calvinistische Wijsbegeerte* vol. 9, no. 1, pp. 35–56.
Strauss, D. (2001) *Paradigms in Mathematics, Physics and Biology: Their Philosophical Roots.* Tekskor, Bloemfontein.
Sullivan, H. (1964) *The Fusion of Psychiatry and Social Science.* New York, Norton.
Suppe, F. ed. (1974) *The Structure of Scientific Theories.* Urbana, University of Illinois Press.
Sweetser, E. (1990) *From Etymology to Pragmatics: Metaphorical and Cultural Aspects of Semantic Structure.* Cambridge, Cambridge University Press.
Tanner, K. (1987) Theology and the Plain Sense. In Green, G. ed. *Scriptural Authority and Narrative Interpretation*, Philadelphia, Fortress Press, pp. 59–78.
Tiger, L, & Fox, R. (1971) *The Imperial Animal.* London, Secker & Warburg.
Torrance, T. (1984) Christian Faith and Physical Science in the Thought of James Clerk Maxwell. In *Transformation and Convergence in the Frame of Knowledge: Explorations in the Interrelations of Scientific and Theological Enterprise*, Grand Rapids Eerdmans, pp. 215–42.
Toulmin, S. (1972) *Human Understanding.* Princeton, Princeton University Press.
Turbayne, C. (1970) *The Myth of Metaphor.* Rev. ed. Columbia, University of South Carolina Press.

Turner, M. (1988) Categories and Analogies. In Helman, D. ed. *Analogical Reasoning: Perspectives of Artificial Intelligence, Cognitive Science, and Philosophy*, Dordrecht, Kluwer Academic Publishers, pp. 25–40.

Tweney, R. (1989a.) Fields of Enterprise: On Michael Faraday's Thought. In Wallace, D &. Gruber, H. eds. *Creative People at Work: Twelve Cognitive Case Studies*, New York, Oxford University Press, pp. 91–106.

Tweney, R. (1989b) A Framework for the Cognitive Psychology of Science. In Gholson, B. Shadish, W. R. Jr., Neimeyer, R., Houts A. *Psychology of Science. Contributions to Metascience.* Cambridge, Cambridge University Press.

Tweney, R. (2006) Toward a Cognitive Historical Understanding of Michael Faraday's Research: Editor's Introduction. *Perspectives on Science,* Volume 14, no.1.

Van Huyssteen, J. (1997) *Essays in Postfoundationalist Theology.* Grand Rapids, Eerdmans.

Van Leeuwen, R. (1990) Liminality and Worldview in Proverbs 1–9. *Semeia* 50, pp. 111–44.

Van Woudenberg, R. (1998) Panmetaphoricism Examined. *Philosophy and Rhetoric* vol. 31, no. 4, pp. 231–47.

Van Hoozer, K. (1988) *Is There a Meaning in This Text? The Bible, the Reader and the Morality of Literary Knowledge.* Grand Rapids, Zondervan.

Varela, F., Thompson, E. & Rosch, E. (1991) *The Embodied Mind: Cognitive Science and Human Experience.* Cambridge, Mass., MIT Press.

Von Bertalanffy, L. (1952) *Problems of Life: An Evaluation of Modern Biological Thought.* London, Watts.

Von Bertalanffy, L. (1972) Response. In Laszlo, E. ed. *The Relevance of General Systems Theory: Papers Presented to Ludwig Von Bertalanffy on His Seventieth Birthday*, New York, George Braziller.

Walsh, B. (1992) *Subversive Christianity: Imaging God in a Dangerous Time.* Bristol, Regius Press.

Weltman, J. (1973) *Systems Theory in International Relations: A Study in Metaphoric Hypertrophy.* Lexington, Heath and Co.

Wertheimer, M. (1938) Numbers and Numerical Concepts in Primitive Peoples. In Ellis, W. ed. A *Source Book of Gestalt Psychology.* New York, Harcourt, Brace & Co.

Williams, L. (1965) *Michael Faraday.* London, Chapman and Hall.

Williams, L. (1973) Kant, Naturphilosophie and Scientific Method. In Giere, R. & Westfall, R. eds. *Foundations of Scientific Method.* Bloomington, Indiana University Press.

Winter, G. (1981) *Liberating Creation: Foundations of Religious Social Ethics.* New York, Crossroad.

Wittgenstein, L. (1958) *Preliminary Studies for the 'Philosophical Investigations' Generally Known as the Blue and Brown Books.* New York, Harper.

Wolters, A. (1985) *Creation Regained: Biblical Basics for a Reformational Worldview.* Grand Rapids, Eerdmans.

Wolters, A. (2000) Confessional Criticism and the Night Visions of Zechariah. In Bartholomew, C., Greene, C., & Möller; K. eds. *Renewing Biblical Interpretation.* Scripture and Hermeneutics Series, Vol.1 Carlisle, UK, Paternoster Press/Grand Rapids, Zondervan, pp. 90–117.

Wolterstorff, N. (1983) *Until Justice and Peace Embrace.* Kampen, Kok.

Wolterstorff, N. (1984) *Reason Within the Bounds of Religion.* 2nd ed. Grand Rapids, Eerdmans.

Woozley, A. (1976) *Theory of Knowledge: An Introduction.* London, Hutchinson.

Wright, N. (2006) *Simply Christian. Why Christianity Makes Sense.* San Francisco, Harper.

Index

FSC
www.fsc.org
MIX
Papier | Fördert gute Waldnutzung
FSC® C083411

Kolibri 360 GmbH
Ferdinand-Jühlke-Straße 7
99095 Erfurt, Deutschland
produktsicherheit@kolibri360.de

Druck:
CPI Druckdienstleistungen GmbH
im Auftrag der
Kolibri 360 GmbH
Ferdinand-Jühlke-Str. 7
99095 Erfurt